# The resilience approach to climate adaptation applied for flood risk

T0186393

Cover:
Wijk aan Zee beach facing coastal defences with Tata Steel in the back
Courtesy of M.H. Gersonius
*One picture, two stories: greenhouse gas emissions lead to climate change and accelerating sea level rise, which in turn reduces the performance of coastal defences*

# The resilience approach to climate adaptation applied for flood risk

**DISSERTATION**

Submitted in fulfillment of the requirements of
the Board for Doctorates of Delft University of Technology
and of
the Academic Board of the UNESCO-IHE Institute for Water Education
for the Degree of DOCTOR
to be defended in public,
on Tuesday, 22 May, 2012 at 12:30 o'clock
in Delft, The Netherlands

by

**Berry GERSONIUS**
Master of Science in Civil Engineering
born in Alkmaar, the Netherlands

This dissertation has been approved by the supervisors:
Prof. dr. C. Zevenbergen
Prof. R.M. Ashley

Members of the Awarding Committee:

| | |
|---|---|
| Chairman | Rector Magnificus, TUDelft |
| Vice-chairman | Rector UNESCO-IHE |
| Prof. dr. C. Zevenbergen | UNESCO-IHE / TUDelft, supervisor |
| Prof. R.M. Ashley | UNESCO-IHE / Univ. of Sheffield, supervisor |
| Prof. dr. Z. Kapelan | University of Exeter |
| Prof. dr. C. Koopmans | VU University Amsterdam |
| Prof. dr. Z.W. Kundzewicz | Polish Academy of Sciences |
| Prof. dr. W.E. Walker | TUDelft |
| Prof. drs. ir. J.K. Vrijling | TUDelft, reserve |

CRC Press/Balkema is an imprint of the Taylor & Francis Group, an informa business

Published by:
CRC Press/Balkema
PO Box 447, 2300 AK Leiden, the Netherlands
e-mail: Pub.NL@taylorandfrancis.com
www.crcpress.com - www.taylorandfrancis.co.uk - www.ba.balkema.nl

ISBN 978-0-415-62485-5 (Taylor & Francis Group)

# Summary

## The (quasi-)stationarity approach

Conventional planning/modification of flood risk management (FRM) systems uses the assumption of (quasi-)stationarity to optimise the engineering system (re)design for future loadings based on the (adjusted) statistical properties of observed (historical) time series of events, such as rainfall intensities or river flows. It assumes that probability density functions (PDF) of future events will be the same as in the recent past, or can be adjusted for non-stationarity (e.g., trends) through statistical analysis in order to obtain PDFs of future events. This approach has worked well in the past, when external drivers were changing at a relatively stable, predictable rate. Traditionally FRM systems have been planned in ways that maintained required performance. Trends due to climate change are, however, more difficult to recognize and predict, making such adjustments more difficult, and future PDFs more uncertain. As an example, climate change scenarios for the Dutch North Sea coast give a sea level rise of 0.35 to 0.60 m for the low scenario in 2100, and of 0.40 to 0.85 m for the high scenario. These uncertain climate change impacts have rendered the (quasi-)stationarity approach as now of limited value for adapting to future change.

## Beyond the (quasi-)stationarity approach

According to the Fourth Assessment Report of the Intergovernmental Panel on Climate Change (IPCC AR4), a number of approaches for climate impact and adaptation assessment are available to succeed the (quasi-)stationarity approach. The IPCC AR4 defines the term "approach" as the main orientation of the climate impact and adaptation assessment, and distinguishes (at least) four approaches: cause-based (or: impact); effect-based (or: vulnerability); top-down; and bottom-up (or: adaptation). Cause-based versus effect-based describes whether the climate impact and adaptation assessment looks forward or backward, respectively, in time from a given reference time. This influences the direction in which the cause and effect chain is followed in the reasoning (e.g., from cause to effect). Top-down versus bottom-up relates to the main orientation of the climate impact and adaptation assessment in terms of spatial scales (e.g., from global to local). This thesis makes a further distinction between the static and the dynamic approach for climate impact and adaptation assessment. Static versus dynamic describes whether the climate impact and adaptation assessment takes a static or dynamic perspective, respectively, on adaptive processes and the effects of these processes at/across different spatio-temporal scales. In this thesis, the dynamic approach is termed the resilience approach—in line with the terminology of the IPCC AR4.

Any one approach (or combination of approaches) for climate impact and adaptation assessment can accommodate a variety of different methods as to how it is delivered. The IPCC AR4 defines the term "method" as a systematic (i.e., stepwise) process of analysis. For example, conventional Net Present Value (NPV) analysis is a frequently used method within the static approach. This method uses a singular climate change scenario to devise a static adaptive strategy, which will determine the investments required. There are unfortunately two major limitations of conventional NPV analysis. Firstly, the method is based on expectations of future investments (assuming e.g. an average or worst-case scenario). There may, however, be other (more extreme) scenarios where the life cycle cost will be different from expectations. Secondly, it uses a deterministic investment path for the adaptive strategy. The working assumption is that the adaptive strategy continues unchanged until the end of the time horizon. This reasoning neglects the effects that management decisions may have under the extremely low or extremely high scenarios, because it assumes commitment by decision makers to a certain investment path. Consequently, the conventional NPV method does not adequately reflect the flexibility that exists in alternative adaptive strategies.

Rather than attempting to devise a static adaptive strategy that requires judgement about which of the various and constantly changing scenarios may be most likely, planners could select a dynamic adaptive strategy. This type of strategy allows for easier adaptation in the future via e.g. incremental adjustments to headroom allowances (i.e., factors of safety). The dynamic adaptive strategy confers the ability, derived from e.g. keeping options open (i.e., in-built flexibility), to adjust to future uncertainties as these unfold. This reduces the effect of decisions made at the start of the adaptation process that might subsequently be found to be not the best, resulting in e.g. unnecessary costs of potentially irreversible measures.

The general objective of this thesis is to investigate the usefulness of a number of different methods within the resilience approach for the development of a dynamic adaptive strategy. A method is considered useful when it provides guidance on when, where and how to adapt to climate (and other) change(s).

**The resilience approach**
The resilience approach for climate impact and adaptation assessment is founded on the understanding that the state of a system is subject to change. It considers adaptation not in the light of specific adaptation options, but rather in how adaptation options feedback, either positively or negatively, into the system as a whole through time and space. Such adaptation options, therefore, need to be conceived as part of a path-dependent trajectory of change. This can be explained as follows: the decisions of the past influence the adaptation options that are available in the

present; and the decisions in the present have implications for the flexibility of which adaptation options can be implemented in the future. The methods within the resilience approach should give insight into these implications. The resilience approach, furthermore, suggests that future change may open up opportunities for incremental adjustments or, possibly, transformational change. The methods within the resilience approach, therefore, need to consider the ability not only to respond to threats (with in-built flexibility), but also to take advantage of opportunities that arise from future change.

## Methods within the resilience approach

This thesis provides (case study) experience with four methods within the resilience approach: Adaptive Policy Making (APM), Real-In-Options (RIO), Adaptation Tipping Point (ATP) and Adaptation Tipping Point - Adaptation Mainstreaming Opportunity (ATP-AMO). These methods are explained below.

APM combines the resilience approach with the so-called risk management framework. Risk management has been defined in the IPCC AR4 as "the culture, processes and structures directed towards realising potential opportunities whilst managing adverse effects". APM deals with change as a threat/opportunity by defining indicators and specific potential adaptations that can be taken in the future once certain thresholds or trigger events are reached. The main limitation of APM is the lack of a clear procedure for the development of a core strategy for maintaining required performance. Rather, it has broader utility as an overarching framework or process for facilitating resilience-focused adaptation. This framework is, therefore, best used in combination with other approaches to develop the core strategy.

RIO analysis combines the resilience approach with the cause-based approach. The caused-based approach begins by considering the changing climate system (drivers), the consequent pressures (e.g., increased runoff), and state (e.g., system performance) to predict the impacts (e.g., flooding and pollution). Responses are then formulated to deal with the pressures and impacts in a way that maintains required levels of performance. RIO analysis uses probabilistic climate data to identify an "optimal" set of static adaptive strategies in response to advances in knowledge about climate change. This involves the estimation of the value of flexibility built into the engineering system (re)design. The value of flexibility stems from the capacity of the decision makers to learn from the arrival of new information and their willingness and ability to revise investment decisions based upon that learning. This is analysed within a framework that builds on (but does not apply) the financial options theory of Black and Scholes. As such, the main benefit of RIO is its ability to deal with change as a threat by explicitly building in flexibility into the engineering system (re)design. A major drawback is, however, that the method assumes probabilities can be given to future loadings under climate change; many climate scientists do not believe this is yet possible. Without probabilities the value of flexibility cannot be estimated. RIO does not provide a procedure for dealing with change as an opportunity, but could potentially be combined with the bottom-up approach to consider this ability.

ATP combines the resilience approach with the effect-based approach. The effect-based approach starts by specifying an outcome (i.e., required performance) used to define acceptability thresholds to manage the impacts, and then assesses the likelihood of attaining or exceeding this outcome as a result of changing drivers. The ATP method examines the effects of increasing design loadings on the system performance. The benefit of ATP is that it is virtually independent of climate change scenarios, and in particular of probabilities of climate change. Climate change becomes relevant for adaptation-related decision making only if it would lead to the crossing of an acceptability threshold. The ATP method, therefore, requires a range of plausible scenarios that can be used to assess whether or not the system is likely to cross any acceptability threshold in the face of climate change. In this sense, the method is more dependent on stakeholder engagement to quantify the acceptability thresholds, to identify the potential options for adapting the system, and to select an adaptive strategy that is realistic and acceptable. ATP can deal with change as a threat by identifying and analysing potential options, or flexibility, for adapting the system to climate change. However, in its simplest form, it lacks a clear procedure for the development of an "optimal" dynamic adaptive strategy. In this respect, the recent extension of the ATP method, called Adaptation Pathways, provides a promising way forward. ATP does not provide a procedure for dealing with change as an opportunity; though it can easily be combined with the bottom-up approach to consider this ability.

ATP-AMO starts with an analysis of ATPs and extends this to include aspects from the bottom-up approach. The extension concerns the analysis of AMOs in the system of interest and other closely related systems. The results from both analyses are then used in combination to take advantage of the right (i.e., cost-efficient) AMOs. ATP-AMO deals with change as a threat in exactly the same way as the ATP method. Its main benefit over ATP lies in its ability to deal with change as an opportunity. The ATP-AMO method provides a well-defined procedure for determining which responses and potential adaptations, where and when to incorporate into 'normal' investment projects, such as for urban regeneration and renewal.

In light of the different approaches for climate impact and adaptation assessment underlying the methods above, it has been concluded that each has particular benefits under particular circumstances. The selection of an appropriate method will depend on a number of factors, including (amongst others): knowledge about the probabilities of climate change; agreement on the potential options for adapting the system; and the capacities and capabilities available on the part of the user(s) of the method.

## Conclusions

The added value of the resilience approach over the static approach derives from the understanding that the state of the FRM system is subject to change. This implies that the degree of system adaptedness to future conditions will change as the system context changes. Adaptedness refers to the effectiveness of the FRM system in meeting the requirements of performance in a specific system state. As an example: the FRM system may be initially designed to deal with the design loading under the medium climate change scenario (i.e., with a high degree of adaptedness), but it may be incapable of adapting to more extreme scenarios. Applying the methods within the resilience approach, e.g. RIO analysis, will provide insight into the trade-offs between the adaptedness of and the flexibility built into the FRM system. In this sense, the use of the resilience approach facilitates the development of responses and potential adaptations that are appropriate at the right time and right cost. The resilience approach, furthermore, suggests that future change, such as that which arises from urban dynamics, may create opportunities for adapting the FRM system to climate change. Application of the methods within the resilience approach, e.g. ATP-AMO, will help to identify and take advantage of the right opportunities. It is, therefore, possible to conclude that the resilience approach has significant potential to support the adaptation of FRM systems to climate change.

# Samenvatting

## De (quasi-)stationaire benadering

Conventionele planning/aanpassing van watersystemen voor de beheersing van overstromingsrisico's en wateroverlast gaat uit van (quasi-)stationariteit om het technisch (her)ontwerp te kunnen optimaliseren voor toekomstige belastingen. Basis hiervoor zijn de statistische eigenschappen van waargenomen (historische) tijdreeksen van gebeurtenissen, zoals neerslagintensiteit of rivierafvoer. De (quasi-) stationariteitsaanname veronderstelt dat de kansdichtheidsfuncties (KDF) van toekomstige gebeurtenissen gelijk zijn aan die in het recente verleden, of dat deze aangepast kunnen worden aan non-stationariteit (bijv. trends) door middel van statistische analyse. Deze aanpak werkte naar behoren in het verleden, toen de veranderingen in externe sturende krachten relatief stabiel en voorspelbaar waren. De watersystemen zijn traditioneel namelijk zo ontworpen dat deze konden blijven functioneren als vereist. Trends als gevolg van de klimaatverandering zijn echter moeilijker te herkennen en voorspellen, wat statistische aanpassing bemoeilijkt en toekomstige KDFs onzekerder maakt. Ter illustratie, klimaatscenario's voor de Nederlandse kust geven een zeespiegelstijging van 0,35 tot 0,65 m voor het lage scenario in 2100, en van 0,40 tot 0,85 m voor het hoge scenario. Deze onzekere klimaateffecten hebben ertoe geleid dat de (quasi-)stationaire benadering vanaf heden van beperkte waarde is voor de planning/aanpassing van watersystemen.

## Voorbij de (quasi-)stationaire benadering

Volgens het Vierde Assessment Rapport van het Intergovernmental Panel on Climate Change (IPCC AR4) is een aantal benaderingen beschikbaar voor het beoordelen van klimaateffecten en adaptatie, die de (quasi-)stationaire benadering zouden kunnen opvolgen. Het IPCC AR4 definieert de term "benadering" als de hoofdrichting van de klimaateffect- en adaptatiebeoordeling, en onderscheidt (op zijn minst) vier benaderingen: oorzaak-gebaseerd (of: impact); gevolg-gebaseerd (of: kwetsbaarheid); top-down; en bottom-up (of: adaptatie). Oorzaak- versus gevolg-gebaseerd beschrijft of de klimaateffect- en adaptatiebeoordeling vooruit respectievelijk achteruit kijkt in de tijd vanuit een bepaald referentietijdstip. Dit beïnvloedt de hoofdrichting waarin de oorzaak en gevolg keten wordt doorlopen (bijv. van oorzaak naar gevolg). Top-down versus bottom-up heeft betrekking op de hoofdrichting van de klimaateffect- en adaptatiebeoordeling in termen van ruimtelijke schaalniveaus (bijv. van mondiaal naar lokaal). Dit proefschrift maakt verder onderscheid tussen de statische en de dynamische benadering voor het beoordelen van klimaateffecten en adaptatie. Statisch versus dynamisch beschrijft of de klimaateffect- en adaptatiebeoordeling een statische respectievelijk dynamische kijk neemt op adaptatieprocessen en de effecten daarvan op/over verschillende ruimtelijk-temporele schaalniveaus. De dynamische benadering wordt in dit proefschrift de veerkrachtbenadering genoemd—conform de terminologie die door het IPCC AR4 gehanteerd wordt.

Elke benadering (of combinatie van benaderingen) voor het beoordelen van klimaateffecten en adaptatie kan een reeks verschillende methodes omvatten met betrekking tot de wijze waarop deze wordt toegepast. Het IPCC AR4 definieert de term "methode" als een systematisch (i.e., stapsgewijs) analyse proces. Netto Contante Waarde (NCW) analyse is bijv. een veel gebruikte methode binnen de statische benadering. Deze methode gebruikt één enkel klimaatscenario om een statische adaptatiestrategie te ontwikkelen, welke de benodigde investeringen bepaalt. NCW analyse heeft echter twee belangrijke beperkingen. Ten eerste is de methode gebaseerd op verwachtingen betreffende toekomstige investeringen (uitgaande van bijv. een gemiddeld of worst-case scenario). Er kunnen echter andere (meer extreme) scenario's optreden waarvoor de investeringen anders zijn dan verwacht. Ten tweede wordt uitgegaan van een deterministisch investeringspad voor de adaptatiestrategie. De aanname is dan dat de adaptatiestrategie ongewijzigd blijft tot aan het eind van de analyse horizon. Deze redeneringswijze negeert het effect van beheersbeslissingen onder extreem lage of extreem hoge scenario's, omdat aangenomen wordt dat besluitvormers vasthouden aan een (vooraf) bepaald investeringspad. Als gevolg geeft de NCW methode geen correcte weergave van de beschikbare flexibiliteit in alternatieve adaptatiestrategieën.

In plaats van een statische adaptatiestrategie, gebaseerd op vooraf bepaalde klimaatscenario's, kunnen planners kiezen voor een dynamische adaptatiestrategie. Een dergelijke strategie biedt gelegenheid verdere adaptatie in toekomst bijv. door incrementele aanpassingen van de voorziene overcapaciteit (i.e., veiligheidsmarges). De dynamische adaptatiestrategie beschikt dus over het vermogen om zich aan te passen aan toekomstige veranderingen. Dit beperkt het effect van eerder genomen beslissingen die achteraf niet de beste blijken te zijn, wat tot onnodige kosten van onomkeerbare maatregelen kan leiden.

De algemene doelstelling van dit proefschrift is om het nut van de reeks verschillende methodes binnen de veerkrachtbenadering te onderzoeken voor de ontwikkeling van een dynamische adaptatiestrategie. Een methode wordt als nuttig beschouwd wanneer deze richting geeft aan de vraag welke adaptatiemaatregelen, waar en wanneer te nemen.

## De veerkrachtbenadering

De veerkrachtbenadering voor het beoordelen van klimaateffecten en adaptatie is gebaseerd op het idee dat de toestand van een system onderhevig is aan verandering. Deze wijze van kijken weerspiegelt hoe specifieke adaptatiemaatregelen door ruimtelijke en temporele terugkoppelingen binnen het gehele systeem met elkaar verbonden zijn. Dergelijke adaptatiemaatregelen moeten daarom als onderdeel van een padafhankelijk veranderingstraject worden beschouwd. Dit kan als volgt worden verklaard: beslissingen uit het verleden beïnvloeden de adaptatiemaatregelen die in het heden beschikbaar zijn, en beslissingen in het heden hebben gevolgen voor de flexibiliteit in maatregelen voor de toekomst. De methodes binnen de veerkrachtbenadering moeten inzicht geven in deze gevolgen. De veerkrachtbenadering gaat er verder vanuit dat toekomstige veranderingen kansen kunnen bieden voor incrementele aanpassingen en transformaties. De methodes binnen de veerkrachtbenadering moeten daarom niet enkel het vermogen om te reageren op bedreigingen (met ingebouwde flexibiliteit) beschouwen, maar ook het vermogen om kansen die voortkomen uit verandering te benutten.

## Methodes om de veerkrachtbenadering toe te passen

Dit proefschrift biedt (praktijk-)ervaring met vier methodes binnen de veerkrachtbenadering: Adaptief Beleid Maken (ABM), Reële-In-Opties (RIO), Adaptatie Knikpunten (AKP), en Adaptatie Knikpunten - Adaptatie Meekoppelmogelijkheden (AKP-AMM). Deze methodes worden hieronder uitgelegd.

ABM combineert de veerkrachtbenadering met het zogenoemde risicobeheerkader. Risicobeheer is door het IPCC AR4 omschreven als "de benodigde cultuur, processen en structuren om potentiële kansen te benutten en tegelijkertijd schadelijke effecten te beheersen". ABM gaat met verandering als bedreiging/kans om door indicatoren en specifieke potentiële adaptatiemaatregelen te formuleren. Deze maatregelen kunnen in de toekomst worden genomen wanneer bepaalde drempelwaarden of trigger gebeurtenissen bereikt worden. De belangrijkste beperking van ABM is het gebrek aan een duidelijke procedure voor de ontwikkeling van een basisstrategie om te blijven functioneren als vereist. Deze methode is eerder nuttig als een overkoepelend kader of proces om veerkrachtgerichte klimaatadaptatie te faciliteren. Dit kader kan dus het best gecombineerd worden met andere benaderingen om de basis strategie te ontwikkelen.

RIO analyse combineert de veerkracht benadering met de oorzaak-gebaseerde benadering. De oorzaak-gebaseerde benadering begint met het beschouwen van het veranderende klimaatsysteem (de sturende factoren), de daaruit voortkomende systeembelastingen (bijv. verhoogde afstroming), en de toestand van het systeem (bijv. het functioneren van het systeem) om zo de effecten te kunnen voorspellen (bijv. overstromingsrisico's en verontreiniging). Dan worden mogelijke ingrepen geformuleerd om met de belastingen en effecten om te gaan zodat het systeem kan blijven functioneren als vereist. RIO analyse gebruikt probabilistische klimaatgegevens om een optimale set statische adaptatiestrategieën te bepalen naar aanleiding van voortschrijdende kennis over de klimaatverandering. Dit betekent dat de flexibiliteit die in het technisch (her)ontwerp is ingebouwd gewaardeerd moet worden. De waarde van flexibiliteit is gebaseerd op het vermogen van de besluitvormer om te leren van nieuw beschikbare informatie en hun bereidheid en mogelijkheid om investeringsbeslissingen aan te passen op basis van de geleerde kennis. Dit kan geanalyseerd worden met behulp van een procedure die voortbouwt op (maar geen gebruik maakt van) de financiële optie theorie van Black en Scholes. Het belangrijkste voordeel van RIO is dus het vermogen om met verandering als bedreiging om te gaan door flexibiliteit in het technisch (her)ontwerp in te bouwen. Een nadeel is echter dat de methode ervan uitgaat dat kansen kunnen worden toegekend aan toekomstige belastingen onder klimaatverandering; veel klimaatwetenschappers denken dat dit nog niet mogelijk is. Zonder kansen kan flexibiliteit niet worden gewaardeerd. RIO biedt geen procedure om met verandering als kans om te gaan, maar kan in potentie worden gecombineerd met de bottom-up benadering om dit vermogen te beschouwen.

AKP combineert de veerkrachtbenadering met de gevolg-gebaseerde benadering. De gevolg-gebaseerde benadering begint met het beschrijven van een uitkomst (i.e., hoe het systeem moet functioneren) die gebruikt wordt om de drempelwaardes voor acceptatie te definiëren om de effecten te beheersen. Vervolgens wordt beoordeeld wat de kans is op het bereiken of overschrijden van deze uitkomst als gevolg van veranderende sturende factoren. Het voordeel van AKP is dat de methode vrijwel onafhankelijk is van klimaatscenario's, en in het bijzonder van kansenverdelingen voor klimaatverandering. Klimaatverandering is enkel van belang voor besluitvorming over adaptatie wanneer deze tot het overschrijden van drempelwaardes voor acceptatie leidt. Voor de AKP methode is daarom een reeks van mogelijke scenario's nodig die gebruikt kunnen worden om te bepalen of er drempelwaardes voor acceptatie worden overschreden onder invloed van klimaatverandering. In dit opzicht is de methode in zekere mate afhankelijk van de inbreng van belanghebbenden om de drempelwaardes te kwantificeren, om de mogelijke adaptatieopties te benoemen, en om een realistische en acceptabele strategie te selecteren. AKP gaat met verandering als bedreiging om door potentiële opties, of flexibiliteit, om het systeem aan te passen aan de klimaatverandering te benoemen en analyseren. Echter, in zijn meest simpele vorm, ontbreekt het aan een procedure om een "optimale" dynamische adaptatiestrategie te ontwikkelen. In dit opzicht biedt de recente uitbreiding op de AKP methode, genaamd Adaptatiepaden, een veelbelovende oplossing. AKP biedt geen procedure om met verandering als kans om te gaan; maar kan eenvoudig worden gecombineerd met de bottom-up benadering om dit vermogen te beschouwen.

AKP-AMM begint met een AKP analyse en voegt daar aspecten van de bottom-up benadering aan toe. De uitbreiding komt neer op een analyse van de adaptatiekansen in het beschouwde systeem met de daaraan gerelateerde systemen. De resultaten van beide analyses worden vervolgens in samenhang beschouwd om de juiste (i.e., kostenefficiënte) AMMs te benutten. AKP-AMM gaat op precies dezelfde wijze met verandering als bedreiging om als de AKP methode. Het belangrijkste voordeel ten opzichte van AKP ligt in het vermogen om om te gaan met verandering als kans. De AKP-AMM methode biedt een uitgewerkte procedure om te bepalen welke ingrepen en potentiële aanpassingen, waar en wanneer mee te nemen in "normale" investeringsprojecten, zoals voor stedelijke herstructurering en vernieuwing.

Gezien de verschillende benaderingen die aan bovenstaande methodes ten grondslag liggen, wordt geconcludeerd dat elke methode bepaalde voordelen heeft onder bepaalde omstandigheden. De keuze van een geschikte methode hangt van een aantal factoren af, zoals (onder meer): kennis over kansenverdelingen voor klimaatverandering; overeenstemming over de potentiële adaptatieopties; en het vermogen en de bekwaamheid van de gebruikers van de methode.

## Conclusies

De toegevoegde waarde van de veerkrachtbenadering boven de statische benadering komt voort uit het besef dat de toestand van het watersysteem onder hevig is aan verandering. Dit betekent dat de mate waarin het systeem is aangepast aan de toekomstige omstandigheden zal veranderen wanneer de systeemcontext verandert. De effectiviteit waarmee het watersysteem voldoet aan de functionele eisen bepaalt de mate van aangepastheid aan bepaalde omstandigheden. Ter illustratie: het watersysteem kan aanvankelijk ontworpen zijn voor een ontwerpbelasting onder het midden klimaatscenario (i.e., met een hoge mate van aangepastheid), maar niet in staat zijn om met meer extreme scenario's om te gaan. Het toepassen van de methodes binnen de veerkrachtbenadering, bijv. RIO analyse, verschaft inzicht in de uitwisselingseffecten tussen de aangepastheid van en de flexibiliteit in het watersysteem. Het gebruik van de veerkrachtbenadering draagt dus bij aan het nemen van de juiste ingrepen en potentiële aanpassingen op het juiste moment en tegen de juiste kosten. Bovendien suggereert de veerkrachtbenadering dat toekomstige verandering, zoals die gerelateerd aan stedelijke dynamiek, kansen biedt voor klimaatadaptatie van watersystemen. Het toepassen van de methodes binnen de veerkrachtbenadering, bijv. AKP-AMM, helpt om de juiste kansen te herkennen en benutten. Daarom kan geconcludeerd worden dat de veerkrachtbenadering een belangrijke bijdrage kan leveren aan de klimaatadaptatie van watersystemen.

# Contents

# Figures and tables

## List of tables

**Table A.1.** Optimal configurations under different approaches to climate adaptation

# Glossary and definition of terms

## Acronyms

**AM**: Annual Maintenance
**AMO**: Adaptation Mainstreaming Opportunity
**APM**: Adaptive Policy Making
**ATP**: Adaptation Tipping Point
**AR4**: Fourth Assessment Report
**cc**: climate change
**CSO**: Combined Sewer Overflow
**Defra**: Department for Environment, Food and Rural Affairs
**DPSIR**: Drivers-Pressures-State-Impacts-Responses
**EA**: Environment Agengy
**e.g.**: exempli gratia, meaning "for example"
**ENPC**: Expected Net Present Cost
**et al.**: et alii, meaning "and others"
**FRM**: Flood Risk Management
**GA**: Genetic Algorithm
**GBM**: Geometric Brownian Motion
**ibid**: ibidem, meaning "the same place"
**IDF**: Intensity-Duration-Frequency
**i.e.**: id est, meaning "that is"
**IenM**: Ministerie van Infrastructuur en Milieu, meaning "Ministry of Infrastructure and the Environment"
**IPCC**: Intergovernmental Panel on Climate Change
**IUD**: Integrated Urban Drainage
**KNMI**: Koninklijk Nederlands Meteorologisch Instituut, meaning "Royal Netherlands Meteorological Institute"
**LAA**: Learning and Action Alliance
**LCC**: Leeds City Council
**MfE**: Ministry for the Environment
**NAP**: Normaal Amsterdams Peil, meaning "Amsterdam Ordnance Datum"
**NBW**: Nationaal Bestuursakkoord Water, meaning "National Water Management Agreement "
**NPC**: Net Present Cost
**NPV**: Net Present Value
**NSGA**: Nondominated Sorting Genetic Algorithm
**PC**: Present Cost
**PDF**: Probability Density Function
**PZH**: Provincie Zuid Holland, meaning "Province of South Holland"

**q**: discharge
**RO**: Real Option
**RIO**: Real In Option
**SES**: Social-Ecological System
**STS**: Socio-Technical System
**SLR**: Sea Level Rise
**SRES**: Special Report on Emissions Scenarios
**SWMM**: Storm Water Management Model
**TAW**: Technische Adviescommissie voor de Waterkeringen, meaning "Technical Advisory Committee on Flood Defences"
**UKCP**: United Kingdom Climate Projections
**USEPA**: United States Environmental Protection Agency
**VenW**: Ministerie van Verkeer en Waterstaat, meaning "Ministry of Transport, Public Works and Water Management"
**WWAP**: World Water Assessment Programme

## Definition of terms

**Acceptability threshold**: The threshold that gives the requirements of performance.

**Adaptation**: The process that entails responding to uncertain changes in drivers, pressures and impacts on a system.

**Adaptability**: The capacity of the actors in a system to manage resilience.

**Adaptation Pathway**: A sequence of responses and potential adaptations, which may be triggered before an ATP occurs.

**Adaptation Tipping Point**: A physical boundary condition where acceptable technical, environmental, societal or economic standards may be compromised.

**Adaptation Mainstreaming Opportunity**: An opportunity for mainstreaming adaptation options with 'normal' investment projects.

**Adaptedness**: The effectiveness of a system in meeting the requirements of performance in a specific system state.

**Adaptive capacity**: See "adaptability".

**Adaptive Policy Making**: A stepwise method for developing adaptive policies taking into account the multiplicity of plausible futures.

**Adaptive potential**: The ability of a system to adapt its structure and processes based on anticipated (re)developments within the assessment period.

**Adjustment**: A process that reduces risk and improves the level of adaptedness of a system.

**Approach**: The main orientation of the climate impact and adaptation assessment.

**Attraction basin**: The part or condition of the system state space that may be thought of as containing a particular attractor toward which the system state tends to go.

**Binomial tree**: A simple representation of the evolution of an uncertain variable.

**Climate change factor**: The ratio between the future and present value of a hydro-climatic variable.

**Climate proofing**: A process aimed at enhancing the resilience of a system or a component of the system to climate change.

**Emergence**: The arising of novel and coherent structures, patterns and properties during the process of self-organization in complex systems.

**Factor of safety**: See "Headroom".

**Flexibility**: The ease or difficulty with which a system or a component of the system can be adjusted to future change.

**Flood Risk Management system**: The whole of the physical systems, actors and rules required to manage flood risk.

**Geometric Brownian Motion**: A continuous-time stochastic process where the logarithm of the uncertain variable follows a random walk (i.e., the Brownian motion).

**Headroom**: The excess capacity added on to the design capacity to allow for future uncertainties that cannot be resolved at the present time; frequently known as a factor of safety.

**Identity**: The minimum of what has to be identified and specified if resilience is to be assessed.

**Identity threshold**: The threshold beyond which the identity of a system changes.

**Indicator**: A parameter or a value derived from a combination of parameters that describes the drivers, pressures, state, impacts or responses.

**Latitude**: The width of an attraction basin. This relates to the maximum amount the system can be changed, before its capacity to recover is compromised.

**Learning Alliance:** A group of individuals or organisations with a shared interest in innovation and the scaling-up of innovation, in a topic of mutual interest.

**Learning and Action Alliance**: See "Learning Alliance". The word Action is added to highlight both the learning and the delivery aspects.

**Mainstreaming, policy-level**: The modification of sector policies and programmes to address climate adaptation

**Mainstreaming, project-level**: The modification of 'normal' investment projects to incorporate adaptation responses.

**Maladaptation**: An action taken supposedly to avoid or reduce vulnerability to climate change that impacts adversely on, or increases the vulnerability of other systems, sectors or social groups.

**Measure, hard structural**: A measure that aims to reduce risks by modifying the system through physical and built interventions.

**Measure, non structural**: A measure that may not require engineering; its contribution to risk reduction is often through changing behaviour through regulation, encouragement and/or economic incentivisation.

**Measure, soft structural**: A measure that involves maintaining or restoring the natural processes with the aim of reducing risks.

**Method**: A systematic (i.e., stepwise) process of analysis.

**Net Present Value**: The sum of the discounted benefits of an alternative less the sum of its discounted costs, all discounted to the same base date.

**Niche**: A network wherein it is possible to deviate from the rules in the existing regime.

**Real Option**: The right—but not the obligation to adjust a system or a component of the system to future uncertainties as these unfold.

**Real In Option**: A Real Option created by changing the engineering system (re)design.

**Regime shift**: The crossing of a social or ecological threshold to another attraction basin.

**Regime, socio-technical**: A relatively stable configuration of institutions, techniques and artefacts, as well as rules, practices and networks that determine the normal development and use of technologies.

**Resistance**: The disturbance required to displace the system by a given amount. This relates to the ease or difficulty of changing the system.

**Resilience, engineering**: The capacity of a system to recover from a disturbance.

**Resilience, ecological/ecosystem**: The capacity of a system to experience shocks while retaining essentially the same function, structure, feedbacks, and therefore identity.

**Resilience, social-ecological**: The capacity of a system to absorb disturbance and reorganize while undergoing change so as to still retain essentially the same function, structure and feedbacks, and therefore identity; that is, the capacity to change in order to maintain the same identity.

**Resilience, socio-technical**: The ability of a system to continue to function as required in the face of change.

**Resilience, technical/infrastructure**: The ability of the technical/infrastructure system to absorb change, so as to continue to function as required in the face of change.

**Resilience approach**: A dynamic perspective on adaptive processes and the effects of these processes at/across different spatio-temporal scales.

**Return period**: The average number of years within which an event is expected to be equalled or exceeded only once.

**Risk management**: The culture, processes and structures directed towards realising potential opportunities whilst managing adverse effects.

**Robustness, decision**: The degree to which a decision or policy performs well under a range of conditions.

**Robustness, dynamic system**: See "Resilience, socio-technical".

**Robustness, system**: See "Resilience, technical/infrastructure".

**Scenario**: Plausible and internally consistent view of the future, which is used to explore uncertain future changes, the potential implications of change and the responses to these.

**Signposts**: Indicators whose development should be tracked in order to determine whether a strategy is meeting its objectives (often translated into acceptable standards).

**Social-Ecological System**: The co-evolutionary units of social and ecological systems.

**Socio-Technical System**: All the physical systems, actors and rules required in order to perform a particular function.

**Stationarity**: The idea that natural systems fluctuate within an unchanging envelope of variability.

**Strategy, adaptive**: A defined set of responses and potential adaptations for maintaining required performance.

**Strategy, dynamic adaptive**: A set of static adaptive strategies with the same initial configuration and different evolutionary configurations. This type of strategy allows for easier adaptation in the future via e.g. incremental adjustments to headroom allowances.

**Strategy, static robust**: A strategy that requires the technical/infrastructure system to be initially designed to accommodate the worst case scenario for future change. This implies the adoption of a headroom methodology.

**Strategy, supporting**: A strategy that addresses the external drivers and internal processes that affect the performance of the core adaptive strategy.

**Trigger**: The critical value of an indicator at which specific potential adaptations are triggered.

**Transformability**: The capacity to create a fundamentally new system when the ecological, economic, or social structures make the existing system untenable.

**Transformational change**: A process that creates a fundamentally new system.

**Uplift**: A specific factor of safety against climate change. An uplift will in general be specific to an emission scenario, climate change model, location and time period.

# 1. Introduction and overview

## 1.1. Introduction

The purpose of this chapter is to describe the (quasi-)stationarity approach to the planning/modification of flood risk management (FRM) systems, to consider why it is necessary to succeed this approach, and to identify a potential way forward for the planning/modification of FRM systems, termed the resilience approach. The objective of this thesis is formulated in the form of a main research question and three sub questions regarding the application of the resilience approach. The methods used to answer these questions are also discussed.

### Conventional planning of FRM systems

Based on the Socio-Technical System (STS) perspective (Geels, 2004), the FRM system is defined, here, as the whole of the physical systems (e.g., flood risk infrastructure), actors (e.g., FRM organisations) and rules (e.g., acceptable flood risk standards) that are required to manage flood risk. The planning/modification of FRM systems requires the definition of a required level of performance. This is typically determined by the frequency of occurrence (i.e., probability) of certain magnitudes of events, such as rainfall intensities or river flows. Or this can relate to risk by defining combinations of probability and consequence. There are major differences in the required performance of the various FRM systems: coastal, river, open water, major and minor drainage. The required performance of coastal, river and open water systems is typically much higher than of major and minor drainage systems (typically 0.01 to 0.001 annual probability compared with 0.1 to 0.01 annual probability, respectively) (e.g., BSI, 1995-1998). The system capacity should be adequate to provide the required level of performance. This (i.e., design capacity) can be defined using probability density functions (PDFs), which describe the frequency of occurrence of different magnitudes of events. The inverse of this frequency is commonly referred to as the return period: a 1 in X year event. A return period of X years corresponds to the average number of years within which an event is expected to be equalled or exceeded only once (WMO, 2009). However, in any time period there is a finite possibility that the event will occur.

One approach to estimating PDFs is via a frequency analysis, which analyses observed (historical) time series of hydro-climatic variables in order to estimate the frequencies of occurrence. Frequency analysis uses either peaks-over-threshold series or annual maximum series. The peaks-over-threshold series contains all events with a magnitude above a specified threshold level, while the annual maximum series contains only the event with the largest magnitude that occurred in each year. When applied to an observed time series, the frequency analysis tests the variables for an assumption of stationarity. Stationarity implies that the variables of the time series should be identically distributed: i.e., they should have the same PDF, which is independent of time (Zevenbergen et al., 2011).

It has long been recognised that external drivers are actually non-stationary and that system dynamics due to physical and socio-economic changes may, in some cases, compromise the stationarity assumption. For instance, the Dutch Delta Report (1960) removed the rise of sea level from the observations before the frequency analysis. Then after extrapolating to a design value for the 1 in 10,000 year event, a sea level rise of 0.20 m/century was added. This approach is, however, still quasi-stationary, because it assumes that the observed time series is stationary with respect to a deterministic trend (e.g., the rise of sea level). For the quasi-stationary hydrology approach, a trend must be recognisable and predictable to allow adjustment of observed time series to future conditions (Olsen et al., 2010). This approach has worked well in the past, when external drivers were changing at a relatively stable, predictable rate. Traditionally FRM systems have been planned in ways that maintained required performance.

Trends due to climate change are, however, more difficult to recognize and predict, making such adjustments more difficult, and future PDFs more uncertain (*ibid*). As an example, climate change scenarios for the Dutch North Sea coast give a SLR of 0.35 to 0.60 m for the low scenario in 2100, and of 0.40 to 0.85 m for the high scenario (Van den Hurk, 2007). These uncertain climate change impacts have rendered the (quasi-)stationarity approach as now of limited value for adapting to future change (Milly et al., 2009; CHS, 2011; WWAP, 2012).

**Beyond the (quasi-)stationarity approach**
According to the Fourth Assessment Report of the Intergovernmental Panel on Climate Change (IPCC AR4) (Carter et al., 2007), a number of approaches for climate impact and adaptation assessment are available to succeed the (quasi-) stationarity approach. The IPCC AR4 defines the term "approach" as the main orientation of the climate impact and adaptation assessment, and distinguishes (at least) four approaches: cause-based (or: impact); effect-based (or: vulnerability); top-down; and bottom-up (or: adaptation) (see Fig. 1.1). Cause-based versus effect-based describes whether the climate impact and adaptation assessment looks forward or backwards, respectively, in time from a given reference time. This influences the direction in which the cause and effect chain is followed in the reasoning (e.g., from cause to effect). Top-down versus bottom-up relates to the main orientation of the climate impact and adaptation assessment in terms of spatial scales (e.g., from global to local) (Jones and Preston, 2011). This thesis makes a further distinction between the static and the dynamic approach for climate im-

pact and adaptation assessment. Static versus dynamic describes whether the climate impact and adaptation assessment takes a static or dynamic perspective, respectively, on adaptive processes and the effects of these processes at/across different spatio-temporal scales. In this thesis, the dynamic approach is termed the resilience approach, based on Nelson et al. (2007) and in line with the terminology of the IPPC AR4.

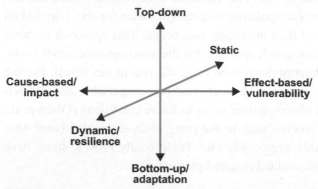

**Figure 1.1.** Approaches for climate impact and adaptation assessment

Any one approach (or combination of approaches) for climate impact and adaptation assessment can accommodate a variety of different methods as to how it is delivered. The IPCC AR4 defines the term "method" as a systematic (i.e., stepwise) process of analysis. The remainder of this section describes the above six approaches in more detail and provides examples of the available methods within each approach. These examples are meant to be illustrative of the approaches, and not to be a thorough review of the methods available within each approach.

Cause-based/impact approach

The cause-based/impact approach begins by considering the changing climate system (drivers) and the consequent pressures (e.g., increased runoff), state (e.g., system performance) to predict the impacts (e.g., flooding and pollution). Responses then need to be formulated to deal with the pressures and impacts in a way that maintains required levels of performance. As an example, it is common to consider adapting FRM systems to climate change by adding simple uplifts to e.g. rainfall intensities or river flows and then assessing whether or not the existing system can cope or not (e.g., Defra, 2006). Such uplifts will in general be specific to an emission scenario, climate change model, location and time period. This is the climate change uplift method, which is similar to (but not the same as) the classical factor of safety method. The climate change uplift method uses a specific factor against climate change only, whereas the classical factor of safety method uses a general factor that addresses a large range of uncertainty, including variables and models; being first formally demonstrated by Rankine in the 1850s and most recently following defined standards for the factor or partial factors of safety (Addis, 2007). The problems associated with e.g. the climate change uplift method stem from the reliance on estimated scenarios that are expected to provide some precision as regards forecasts of climate change. However, despite past and current scientific advances in climate modelling, there remain large uncertainties about the direction, rate and magnitude of climate change. Uncertainty associated with climate modelling arises from model errors, internal variability and emissions scenario uncertainty (Cox and Stephenson, 2007). Whilst climate science can potentially reduce the uncertainty from model errors and, to some extent, also from internal variability, this uncertainty reduction will be a gradual and lengthy process and in itself assumes some quasi-stationarity, i.e., that there will not be any sudden change in drivers. Nevertheless there will always be significant irreducible uncertainty related to future emissions and consequent climate changes; this has been referred to as deep uncertainty (Lempert and Schlesinger, 2000). Additionally, there is uncertainty about how global climate changes will influence changes in hydrological processes especially at the regional scale (Willems et al., 2011). These climate change uncertainties will limit the usefulness of the cause-based/impact approach for adaptation-related decision making. This is because an uncertainty cascade arises when climate change uncertainties are applied to impact models. This concerns the process whereby many of the uncertainties from each step of the assessment accumulate; resulting in large ranges of possible impacts (Schneider, 1983). Such ranges commonly become too large for practical application in planning.

Effect-based/vulnerability approach

The effect-based/vulnerability approach starts by specifying an outcome (i.e. required performance) used to define acceptability thresholds to manage the impacts, and then assess the likelihood of attaining or exceeding this outcome as a result of changing drivers (Lempert et al., 2004). An example of this is the exploratory modelling-based method for robust adaptation decision making (Lempert et al., 2003). This uses computer modelling to develop a large ensemble of future scenarios, where each scenario represents one possible set of boundary conditions as well as one possible choice among many alternative adaptive strategies. It aims to identify adaptive strategies that are robust under a wide range of future scenarios.

Top-down approach

The top-down approach considers the outputs of global climate models, which are downscaled to regional climate models to serve as input to hydrological models to assess impacts (Parry and Carter, 1998). Adaptive strategies are then developed based on the likely physical impacts of climate change on the system of interest. However, as a consequence, such an approach tends to neglect the wider contexts—including spatial planning, economic priorities, technical regulation, cultural preferences, risk psychology, etc.—in which adaptation has to take place (Dessai et al., 2009).

## Bottom-up/adaptation approach

As many characteristics of adaptation tend to be location-specific, there is currently an increasing recognition of the bottom-up/adaptation approach. This type of approach commences at the local scale, assessing the existing system to determine whether it is feasible to increase its ability to deal with climate change, including the variability (Jones and Boer, 2005). It also takes account of climate model predictions for the assessment of robust adaptation requirements through scenario-based approaches (e.g., Evans et al., 2004). This approach is based on the recognition that adaptation is better conceived as a socio-economic process rather than as a set of stand-alone adjustments, taking a more dynamic view of adaptation by combining climate change with socio-economic drivers (Jones and Preston, 2011). This has also been referred to as 'adaptation mainstreaming' (Huq and Reid, 2004). According to Persson and Klein (2009), there is an important distinction between adaptation mainstreaming at the policy/programme-level (e.g., Huq and Reid, 2004) and the project-level (e.g., Zevenbergen et al., 2007; Veerbeek et al., 2010). The former has to do with the modification of sector policies and programmes to address climate adaptation, while the latter concerns the modification of 'normal' investment projects to incorporate adaptation responses. Project-level adaptation mainstreaming is the most relevant level for the development of adaptive strategies.

## Static approach

The large ranges of possible climate impacts, due to the uncertainty cascade (Schneider, 1983), have frequently led to the pitfall that a singular climate change scenario is adopted by policy makers, planners or others as an average or worst-case to be prepared for. In this case, the results of the climate impact and adaptation assessment will be highly dependent on the chosen scenario and the assumptions concerning the related uncertainties (Kwadijk et al., 2010). This is the static approach, which has been termed Predict-Then-Adapt (Hulme, 2009). The methods within the static approach are decoupled from climate change uncertainties and the resulting adaptive strategy is, therefore, static (i.e., inflexible). As an example, conventional Net Present Value (NPV) analysis uses a singular climate change scenario to devise a static adaptive strategy, which will determine the investments required. There are unfortunately two major limitations of conventional NPV analysis. Firstly, the method is based on expectations of future investments (assuming e.g. an average or worst-case scenario). There may, however, be other (more extreme) scenarios where the life cycle cost will be different from expectations. Secondly, it uses a deterministic investment path for the static adaptive strategy. The working assumption is that the adaptive strategy continues unchanged until the end of the time horizon. This reasoning neglects the effects that management decisions may have under the extremely low or extremely high scenarios, because it assumes commitment by decision makers to a certain investment path. Consequently, the conventional NPV method does not adequately reflect the flexibility that exists in alternative adaptive strategies.

Dynamic/resilience approach

The dynamic/resilience approach is founded on the understanding that the state of a system is subject to change. It considers adaptation not in the light of specific adaptation options, but rather in how adaptation options feedback, either positively or negatively, into the system as a whole through time and space (Nelson et al., 2007; Zevenbergen et al., 2008). Such adaptation options, therefore, need to be conceived as part of a path-dependent trajectory of change. This can be explained as follows: the decisions of the past influence the adaptation options that are available in the present; and the decisions in the present have implications for the flexibility of which adaptation options can be implemented in the future (*ibid*). The methods within the dynamic/resilience approach should give insight into these implications. The dynamic/resilience approach, furthermore, suggests that future change may open up opportunities for incremental adjustments or, possibly, transformational change (Folke et al., 2010). The methods within the dynamic/resilience approach, therefore, need to consider the ability not only to respond to threats (with in-built flexibility), but also to take advantage of opportunities that arise from future change (Nelson et al., 2007). Within the dynamic/resilience approach, Decision Analysis has been used to assess the value of flexibility (De Bruin and Ansink, 2011). The value of flexibility stems from the capacity of the decision makers to learn from the arrival of new information and their willingness and ability to revise investment decisions based upon that learning. Decision Analysis structures the adaptation options into a decision tree, distinguishing event nodes (that represent uncertain outcomes with attached subjective probabilities) and decision nodes (that represent choices by the decision maker). The decision rule is to identify the strategy that provides the best expected value, as a weighted average of the outcomes by their probability of occurrence (de Neufville, 1990). A more advanced method in determining the value of flexibility is Real Options (RO) analysis (Myers 1984). RO analysis estimates the value of flexibility within a framework that builds on (but does not apply) the financial options theory of Black and Scholes (1973).[1] Although the RO analysis is (theoretically) superior to Decision Analysis in determining the value of flexibility, its implementation requires probabilistic climate change data (which is usually not available).

## 1.2. Objective and research questions

The previous section has described how, along with a number of external drivers (e.g., climate change), the factors influencing the FRM system have changed from a relatively stable, predictable system with only slowly changing external drivers to a less predictable system subject to a lack of stationarity. This increasing lack of stationarity, and of the (un)predictability of loading and effects, makes it necessary to succeed the (quasi-)stationarity approach. A key requisite is, therefore, to identify and/or update approaches and methods that can be used to deal with non-stationarity induced by climate (and other) change(s).

A frequently used approach to deal with climate change impacts and adaptation is the static approach (or: Predict-Then-Adapt). This approach uses a singular climate change scenario to devise a static adaptive strategy. Because the future cannot be predicted (e.g., Cox and Stephenson, 2007), this strategy might subsequently be found to be not the best. Therefore, rather than attempting to devise a static adaptive strategy that requires judgement on which of the various and constantly changing scenarios may be most likely, planners could select a dynamic adaptive strategy (Walker et al., 2001). This type of strategy allows for easier adaptation in the future via e.g. incremental adjustments to headroom allowances (i.e., factors of safety). The dynamic adaptive strategy confers the ability, derived from e.g. keeping options open (i.e., in-built flexibility), to adjust to future uncertainties as these unfold. This reduces the effect of decisions made at the start of the adaptation process that might subsequently be found to be not the best, resulting in e.g. unnecessary costs of potentially irreversible measures. A portfolio of structural and non-structural measures is usually, though not necessarily, required for the implementation of the dynamic adaptive strategy to ensure that cost-effective adaptation can take place in all future time periods. Non-structural measures correspond to the design and application of policies and procedures, and employing among other land-use controls, information dissemination, and economic incentives to reduce risks (EC, 2009).

---

[1] In 1973 Black, Scholes and Merton (Black and Scholes, 1973; Merton, 1973) determined a closed form solution to value simple put and call options, given assumptions about the behaviour of the underlying asset.

The saw-tooth effect in probability/risk with time, as a consequence of taking a dynamic adaptive strategy, is represented in Fig. 1.2. This diagram shows the probability/risk increasing with time, together with the acceptable standard. The acceptable standard may be defined either based on the likelihood of flooding or based on a broader risk-based approach, taking account of the likelihood as well as consequences. Under the risk-based approach the acceptable standard will be the economically optimal level of flood risk in terms of costs and benefits. The focus of this PhD thesis is predominantly on the likelihood-based approach; although the results are equally valid for the risk-based approach. It is, furthermore, of note that the acceptable standard for FRM is required to keep pace with the external change drivers, and, therefore, may not be represented properly by the single horizontal line in Fig 1.2. However, it may alter either up or down.

The vertical lines in Fig. 1.2 show the responses and potential adaptations. The difficulty is to decide when these are required and likely to be cost-effective as part of a dynamic adaptive strategy (Ingham et al., 2006) and this question is addressed in this thesis.

The general objective of this thesis is to investigate the usefulness of a number of different methods within the resilience approach for the development of a dynamic adaptive strategy. A method is considered useful when it provides guidance on when, where and how to adapt in relation to the diagram in Fig. 1.2.

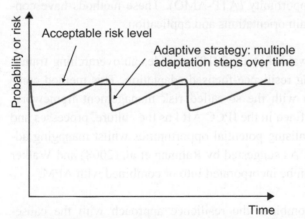

**Figure 1.2.** Graph of probability/risk with time as a consequence of taking a dynamic adaptive strategy

The main research question is:

> **Can the resilience approach support the adaptation of FRM systems to climate change?**

The following sub questions are derived from the main research question to guide the research:

1. How can resilience, and closely related terms, be defined and assessed for STS?
2. Which methods can be used within the resilience approach for climate impact and adaptation assessment? What are the benefits and limitations of the different methods?
3. What is the added value of the resilience approach for FRM?

## 1.3. Methods used

Various methods have been used to answer each sub question above. Relevant literature on the concept of resilience was studied to gain an understanding of its diverse interpretations and applications. From this literature study, an approach emerged, based on the concept of identity (Cumming et al. 2005) that forms the foundation for understanding and applying resilience with respect to STS. This approach has been demonstrated using the example of FRM for the Island of Dordrecht (the Netherlands) (sub question 1).

Following the specific understanding of resilience, a further literature study was conducted on the methods that can be used within the resilience approach. Four methods have been examined in detail: Adaptive Policy Making (APM), Real-In-Options (RIO), Adaptation Tipping Point (ATP) and Adaptation Tipping Point - Adaptation Mainstreaming Opportunity (ATP-AMO). These methods have considerable differences in e.g. main orientations and application.

APM (Walker et al., 2001; Kwakkel et al., 2010) provides an overarching framework or process for facilitating resilience-focused adaptation. This method combines the resilience approach with the so-called risk management framework. Risk management has been defined in the IPCC AR4 as the culture, processes and structures directed towards realising potential opportunities whilst managing adverse effects (AS/NZS, 2004). As suggested by Rahman et al. (2008) and Walker et al. (2012), other methods can be incorporated into or combined with APM.

RIO (De Neufville, 2003) combines the resilience approach with the cause-based/impact approach. It uses probabilistic climate data to identify an "optimal" set of static adaptive strategies in response to advances in knowledge about climate change. This involves the estimation of the value of flexibility built into the engineering system (re)design. RIO analysis embeds the Real Options directly into the engineering system (re)design, which requires extensive knowledge about the technical/infrastructure system.

ATP (Kwadijk et al., 2010) combines the resilience approach with the effect-based/vulnerability approach. The ATP method is aimed at assessing whether, and for how long, the performance of the existing system will continue to be acceptable under different climate conditions. It uses the concept of ATPs, which are reached if the magnitude of climate change is such that acceptable technical, environmental, societal or economic standards may be compromised (Haasnoot et al., 2009).

ATP-AMO starts with an analysis of ATPs and extends this to include aspects from the bottom-up approach. The extension concerns the analysis of AMOs in the system of interest and other closely related systems. The results from both analyses are then used in combination to take advantage of the right (i.e., cost-efficient) AMOs.

A larger range of methods can be applied within the resilience approach, but these were selected in order to cover a range of different approaches in combination with the resilience approach, as shown in Fig. 1.3. Other methods include, but are not limited to: Decision Analysis; Adaptation Pathways; and Adaptation Policy Pathways. Adaptation Pathways (Haasnoot et al., 2012) has combined the ATP method with exploratory modelling. This method sets out to explore a range of relevant adaptation options before an ATP has been reached in order to develop different Adaptation Pathways. An Adaptation Pathway refers to a sequence of responses and potential adaptations, which may be triggered before an ATP occurs (*ibid*). Adaptation Policy Pathways (Walker et al., 2012) incorporates Adaptation Pathways into APM.

APM has been applied to the modification of an urban drainage system in West Garforth, Yorkshire (England). RIO analysis has been developed and demonstrated for the same case study. It has, furthermore, been applied to the semi-hypothetical example of a coastal defence system at the Dutch North Sea coast (the Netherlands). The ATP-AMO method has been developed and demonstrated for the management of flood risk for an urban drainage system in Dordrecht (the Netherlands). Fig. 1.4 shows the locations of the case studies in the North Sea Region. The case study applications have provided insights into the benefits and limitations of each method. These insights were used to compare the various methods and to give specific recommendations as to which method to use under what circumstances (sub question 2).

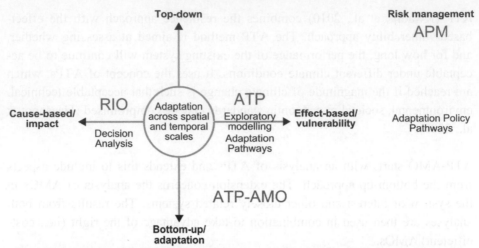

Figure 1.3. Methods within the dynamic/resilience approach mapped against the other approaches: cause-based/impact; effect-based/vulnerability; top-down; bottom-up/adaptation; and risk management (adapted from Jones and Preston, 2011). Risk management has not been included in the diagram, because it does not relate to a main orientation; rather, it provides an overarching framework. The methods selected in this thesis are shown in red

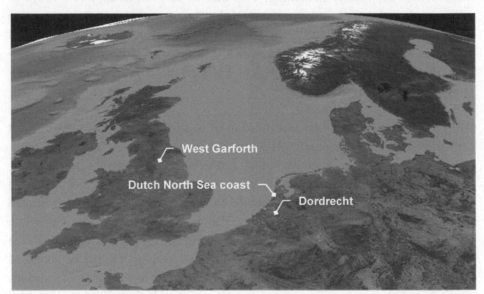

Figure 1.4. Location of the case studies in the North Sea Region (source: Google Earth 2011)

The added value of applying the resilience approach for FRM was determined by comparing the approach with the sequence of other approaches used (through time) to deal with changing flood risk (sub question 3).

## 1.4. Definitions used

There are a number of terms in this thesis (e.g., identity, resilience, robustness) that are open to debate. Therefore, a definition of these terms is necessary at this point to establish a conceptual foundation for the research.

The term resilience has been defined in literature in at least three major ways, from its more narrow interpretation to the broader meaning in relation to social-ecological systems (SES) (Folke, 2006). Table 1.1 summarizes the literature review of Folke (2006).

**Table 1.1.** Major definitions of resilience (source: Folke, 2006)

| Definitions | Characteristics | Focus on | Context |
|---|---|---|---|
| Engineering resilience | Return time, efficiency | Recovery, constancy | Vicinity of a stable equilibrium |
| Ecological/ecosystem resilience, social resilience | Buffer capacity, withstand shock, maintain function | Persistence, robustness | Multiple equilibria, stability landscapes |
| Social-ecological resilience | Interplay disturbance and reorganization, sustaining and developing | Adaptive capacity, transformability, learning, innovation | Integrated system feedback, cross-scale dynamic interactions |

Engineering resilience (Holling, 1996) refers to the dynamics of a system close to a stable equilibrium. This interpretation is concerned with the constancy of state within the basin of attraction, and can be assessed by the speed of return to equilibrium following a disturbance. It is often addressed in terms of recovery capacity. Engineering resilience is a frequently used concept with respect to FRM. As an example, De Bruijn (2005) uses engineering resilience in a study on lowland river systems, which has defined resilience as "the ability of a system to recover from floods in the area". This thesis, however, does not deal with engineering resilience as such.[2]

Because of the existence of multiple equilibria, return time does not measure all of the ways in which a system may fail to maintain its functions. Ecological/ecosystem resilience (Holling, 1973) refers to the ability of a multi-stable sys-

---

[2] This does not mean that the definition of De Bruijn (2005) has been rejected; on the contrary, engineering resilience as applied by De Bruijn is a very useful concept for FRM.

tem to keep the values of its state variables within a given basin of attraction in the face of change. It can be assessed by the magnitude of the disturbance that can be absorbed before the state of the system falls outside its basin of attraction. Ecological/ecosystem resilience often refers to the buffer capacity that allows persistence, or the capacity to absorb disturbance. Resilience as persistence has been defined as: "the magnitude of disturbance that can be absorbed before the system changes its structure by changing the variables and processes that control behavior" (Gunderson and Holling, 2002); and: "the capacity of a system to experience shocks while retaining essentially the same function, structure, feedbacks, and therefore identity" (Walker et al., 2006).

More recently, resilience has increasingly been applied to linked social-ecological systems (SES). The reason for extending the use of resilience to SES is that any delineation between social and ecological systems is seen as artificial and arbitrary (Berkes and Folke, 2000). Social-ecological resilience has been defined as: "the capacity of a system to absorb disturbance and reorganize while undergoing change so as to still retain essentially the same function, structure and feedbacks, and therefore identity, that is, the capacity to change in order to maintain the same identity" (Folke et al., 2010). This definition has extended the meaning of resilience beyond (just) persistence. It incorporates the dynamic interplay of resilience as persistence, adaptability and transformability. Adaptability relates to the capacity of a system to learn, adjust its responses to changing external drivers and internal processes, and continue development along the current trajectory (Berkes et al., 2003). Adaptability has been defined as "the capacity of actors in a system to manage resilience" (Walker et al., 2004). The collective capacity of actors to do this, through purposeful adjustments, determines whether they can successfully avoid crossing social or ecological thresholds. Transformability, by contrast, is the capacity to cross thresholds into new development trajectories (Folke et al., 2010). Transformability has been defined as "the capacity to create a fundamentally new system when ecological, economic, or social structures make the existing system untenable" (Walker et al., 2004).

Robustness is a term that is closely related to resilience. At least two major definitions of robustness can be distinguished: system robustness and decision robustness. System robustness is similar to resilience as persistence, and is more relevant when applied to technical/infrastructure systems (e.g., Anderies et al., 2004; Mens et al., 2011). Decision robustness has been defined as: "the degree to which a decision or policy performs well under a range of conditions" (Lempert et al., 2003).

Resilience as persistence (i.e., system robustness), adaptability and transformability determine the dynamics of a SES (i.e., its future trajectories). There is, however, an important distinction between resilience as persistence and adaptability, on the one hand, and transformability on the other. Resilience as persistence and adaptability are concerned with the dynamics of a particular system or a closely related set of systems; while transformability has to do with fundamentally altering the nature of a system so that it becomes a different kind of system. Given that the distinction between 'a closely related set of systems' and 'a different kind of system' can be fuzzy (Walker et al., 2004), there is no clear dividing line between adaptation conceived as an incremental adjustment and as a transformational change (Nelson et al., 2007).

In this research, the concept of identity comprises four aspects that constitute the minimum of what has to be identified and specified if resilience is to be assessed (Brand, 2009). These are: (a) components, which include the structural variables (both technical and social) that make up the system; (b) relationships, which are the process or interaction variables that link the components; (c) innovation, which are the variables that generate change of components and relations; and (d) continuity, which describes the variables that facilitate the continuation of components and relations through time (Cumming et al., 2005). The identity definition is used e.g. in psychology to study human resilience. Human resilience is determined by the capacity of individuals, communities and institutions to maintain their identity while undergoing transformations through persistence with normal functions and rituals which make them who they are as people, communities and/or institutions (Walker, 2010).

The above definitions from social-ecological studies are applied in this thesis with respect to STSs. These link physical systems with actors and rules in order to provide a particular function (Geels, 2004).

## 1.5. Outline of this thesis

This thesis is comprised of eight chapters. Chapter 1 comprises this introduction.

Chapter 2 provides a definition of resilience with respect to STS. It describes the identity approach for defining and assessing resilience and illustrates this approach for FRM for the Island of Dordrecht (the Netherlands).

The next five chapters (3, 4, 5, 6 and 7) deal with a number of different methods that can be used within the resilience approach. Chapter 3 presents the development of the process for facilitating resilience-focused adaptation: i.e., the APM method. Chapters 4 and 5 provide experience with RIO analysis for the context of an urban drainage system and a coastal defence system, respectively. In Chapter 6, the ATP method is further developed into the ATP-AMO method and then applied to an urban drainage system. Chapter 7 compares RIO and ATP and gives specific recommendations on which method to use under what circumstances.

Chapter 8 combines the results of this thesis to answer the research questions that, in Chapter 1, were derived from the general objective. The conclusions are split into conclusions regarding the sub questions and a main conclusion. Recommendations for practice and further scientific research for advancing resilience-focused adaptation are also given in this chapter.

The body of this thesis (Chapters 2-7) consists of six papers that are under review, have been accepted, or have already appeared in peer-reviewed journals. The intention of each paper is that it is sufficiently self-contained, so as to be understandable without recourse to the other papers. As a result, there is some overlap in content between the various papers, particularly regarding the approaches and methods used and in some cases results are repeated.

# 2. Definition and assessment of resilience for socio-technical systems

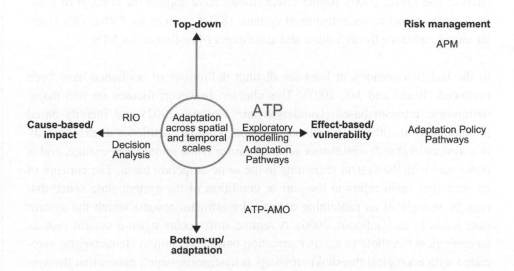

This chapter is based on:
Gersonius, B., Ashley, R. & Zevenbergen, C. 2012. The identity approach for assessing socio-technical resilience to climate change: example of flood risk management for the Island of Dordrecht. *Natural Hazards and Earth System Sciences*, under review.

## 2.1. Introduction

Enhancing resilience to climate change (or: climate proofing (Kabat et al., 2005)), is developing as a best practice concept in relation to flood risk management (FRM) (e.g., European Commission, 2009). The definition of resilience is, however, open to debate and this makes it difficult to apply in practice. Resilience is a concept originally developed for ecological systems (Holling, 1973), and has increasingly been used in many studies on social-ecological systems (SES) (Folke, 2006). The reason for extending the use of resilience to SES is that any delineation between social and ecological systems is seen as artificial and arbitrary (Berkes and Folke, 2000). Rather fewer studies have applied the concept of resilience with respect to socio-technical systems (STS), such as for FRM. This chapter aims to advance the definition and assessment of resilience for STS.

In the last two decades, at least ten distinct definitions of resilience have been produced (Brand and Jax, 2007). This chapter, however, focuses on two major definitions: attractor-based (Gunderson and Holling, 2002) and identity-based (Cumming et al., 2005). The attractor approach defines resilience as the capacity of a system to absorb disturbance and reorganize while undergoing change, and is concerned with the system remaining in the same attraction basin. The concept of an attraction basin refers to the part or condition of the system state space that may be thought of as containing a particular attractor toward which the system state tends to go (Gallopín, 2006). A regime shift occurs when a system crosses an ecological threshold to another attraction basin. A common characteristic associated with ecological threshold crossings is hysteresis, which means that the system change may be irreversible once a threshold has been crossed, even if the driving force that initiated the threshold crossing ceases (Scheffer et al., 2001). Such a regime shift represents a loss of resilience of the system. The degree of resilience is thus quantified by the magnitude of disturbance that a system can undergo before crossing the limit of the attraction basin (Carpenter et al., 2001). Examples of attractor-based resilience assessment are given in Scheffer et al. (2001).

The identity approach equates resilience with the ability of a system to maintain its identity in the face of change, and investigates whether or not the system crosses any key identity thresholds. The concept of identity comprises four aspects that constitute the minimum of what has to be identified and specified if resilience is to be assessed (Brand, 2005). These are: (a) components, which include the structural and non-structural variables that make up the system; (b) relationships, which are the process or interaction variables that link the components; (c) innovation, which are the variables that generate change of components and relations; and (d) continuity, which describes the variables that facilitate the continuation of components and relations through time (Cumming et al., 2005). The rationale behind the identity approach is that many variables within the system may change over time, but the specific variables that define its identity must be maintained if the system is resilient (*ibid*). As such, the degree of resilience is estimated by the potential for a change in identity (and its magnitude) under alternative scenarios for external drivers.

While the attractor approach is useful for assessing resilience of SES, it has an important limitation when applied to STS. This is because STS do not exhibit ecological thresholds, but acceptability thresholds. These set out the requirements of system performance. Here, it is contended that required performance is best conceptualised in terms of maintaining identity in order to ensure that the STS has adequate resilience—rather than considering a set of attractors.[3] The reason for this is that crossing acceptability thresholds does not lead to irreversible system changes; it simply means that a change in the adaptive strategy is required to restore the system performance to its original identity. The concept of identity is, therefore, used as a vehicle to advance the definition and assessment of resilience for STS. The application of the identity approach is illustrated using the example of FRM for the Island of Dordrecht, the Netherlands.

## 2.2. Method

This chapter considers resilience as a quantitative and measurable concept. When used in this sense, it is necessary to specify resilience "to what" (Step 1) and "of what" (Step 2) and, subsequently, to undertake an assessment of the system resilience (Step 3) (Walker et al., 2002). Resilience assessment based on identity is comprised of three sub-steps (Cumming et al., 2005). Step 3A is to develop a conceptual model. Part of this step is to determine the boundaries, such as the spatial and temporal scales of the resilience assessment. The conceptual model is used in Step 3B to identify the specific variables and threshold values that reflect changes in identity. Finally, step 3C assesses the potential for changes in identity under the drivers specified in Step 1. A flow chart of the identity approach for assessing socio-technical resilience is presented in Fig. 2.1. These steps are explained below for the context of FRM.

---

[3] It can be argued that the attractor approach can still be applied metaphorically to the adaptation of STS. However, this is viewed here as being against the spirit of the attractor approach, which focuses on avoiding the crossing of non-returnable thresholds (e.g., Scheffer and Carpenter, 2003).

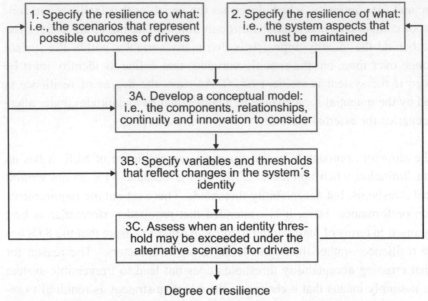

Degree of resilience

**Figure 2.1.** Flow chart of the identity approach for assessing resilience of STS

### Step 1: Resilience to what?

The "to what part" specifies the variables that cause change to the FRM system (i.e., the kind of drivers), with particular relevance to their impacts on the object(s) of interest. Because this chapter deals exclusively with climate proofing, for simplicity this is taken as the single driver of interest for the resilience assessment.[4] Climate change is expected to result in significant changes in the frequency and risk of flooding in many regions. The Intergovernmental Panel on Climate Change (IPCC) has developed different scenarios of climate change. The SRES scenarios used in the IPCC Third Assessment Report were based on likely greenhouse gas emissions in each scenario, together with an assessment of the likely management of these emissions. More recently, the IPCC has provided estimates of the ranges within which global climate changes may occur, given as probabilities (Solomon et al., 2009). Whilst these are useful for the purpose of resilience assessment, these changes need to be considered in terms of the more regional/local impacts on the FRM system.

### Step 2: Resilience of what?

The "of what part" characterises what is being considered as the object(s) of resilience. A critical question in this regard is whether the object of resilience is structural or functional (Smith and Stirling, 2010). Definitions of resilience that make

---

[4] A fuller study of resilience should also take account of socio-economic drivers, which are generally forcing a quicker rate of change on society than the climate.

no distinction between structure and function can become problematic, in particular for STS. This is because resilient individual structures at particular scales (e.g., large-scale engineering structures or tightly regulated institutions) can, in some cases, threaten the performance of the function provided by the STS. The aim of resilience management is, therefore, to enhance or maintain the performance of the function of interest and also to preserve those structures (both technical and social) that lead to enhanced or the same performance—and not necessarily preserve the existing systems themselves (*ibid*). Folke et al. (2010) conclude that sometimes transformations are necessary to reduce this structural resilience in order to gain functional resilience under changed conditions. As an example: the transformation from a hard coastal defence system (e.g., a single sea dike) to an integrated hard/soft coastal defence system (e.g., a sea dike with an elevated, sandy foreshore) will help to deliver increased flexibility to respond to future uncertainties associated with sea level rise, and could, therefore, potentially enhance functional resilience (see Chapter 5).

## Step 3A: Development of a conceptual model

The FRM system has been defined, in this thesis, as the whole of the physical systems, actors and rules that are required to manage flood risk. Therefore, aspects of identity (components, relationships, innovation and continuity) have been selected that relate directly to this function. These are summarised in the next sections and in Table 2.1.

**Table 2.1.** Aspects of Identity

|  | Variables | Explanation (not exhaustive) |
|---|---|---|
| **Structural components** | Physical FRM system | Engineering structures and the environment |
|  | Actors | Individuals, groups and flood management organisations |
|  | Regulative rules | Laws, regulations, policies, procedures and standards |
|  | Normative rules | Values and norms |
|  | Cognitive rules | Belief systems and expectations |
| **Functional relationships** | Causal relations | Drivers/pressures act upon the system state to create the impacts, to which actors will develop responses |
|  | Normative relations | The physical FRM system and actors are structured by rules |
| **Innovation** | Niche dynamics | Articulation and refinement of visions, learning processes and build-up of social networks |
| **Continuity** | Regime dynamics | Institutional common-sense, regulations and standards, adaptation of lifestyles to technical systems, and sunk investments in infrastructures and competencies |

Components

The FRM system encompasses not only the physical FRM system, but also the actors that are impacted upon by flooding or responding to flood risk and includes the rules which structure and regulate the associated physical and socio-economic processes. The physical FRM system is comprised of both the engineering structures provided to deal with flood risk and the environment. The actors involved in the FRM system are individuals, groups, and FRM organisations. Rules can be categorised as formal, normative and cognitive (i.e., regulations, behavioural norms and knowledge, respectively) (Geels, 2004). Examples of formal rules are regulations, laws, procedures and standards; examples of normative rules are values and norms; and examples of cognitive rules are shared belief systems and expectations. Actors in different groups share different kinds of rules, which are referred to as socio-technical regimes (*ibid*). As an example, different groups have different expectations toward flood risk. Rules are not just shared in or between groups, but can also be embedded in the practice of providing engineering structures and of how flood risk is managed. The FRM system is continuously changing because of physical and socio-economic processes (i.e., the drivers), as well as the responses intended to reduce the risk of flooding (Hall et al., 2003). Fig. 2.2 gives a simple conceptualisation of the FRM system with its different components in mutual interaction.

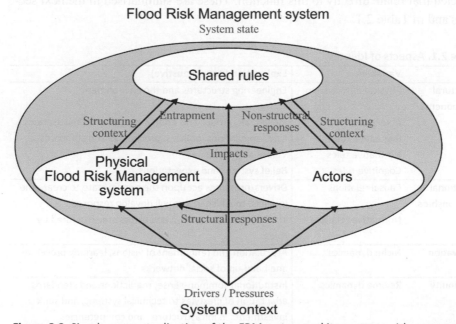

**Figure 2.2.** Simple conceptualisation of the FRM system and its context, with components in oval boxes and relations in arrows

Functional relationships

Different relations exist between the system components and between the system and its context (Ottens et al., 2006). These describe how the system components interact or fit together. The causal relations within the FRM system may be considered to be well-understood, and these are generally described using the Drivers-Pressures-State-Impacts-Responses (DPSIR) model (Evans et al., 2004). The state of the FRM system includes the state of the physical FRM system, actors and rules. In any system state, the FRM system has a quasi-stationary level of risk associated with it, where risk is considered as a function of the flood frequencies and impacts. Drivers, pressures and impacts are then considered in terms of how the system state may alter. Drivers and pressures act upon the system state, often resulting in physical and socio-economic changes. This has both negative and positive effects on the level of risk associated with the system state, and this is described by the impacts. The drivers, pressures and impacts may lead to responses, which are diverse adaptations to the structures and processes by the actors. These can be categorised as either structural or non-structural (European Commission, 2009). Structural responses are engineering-based adaptations to reduce flood risk. Non-structural responses may not require engineering and their contribution to risk reduction is most likely through changing behaviour through regulation, encouragement and/or economic incentivisation (Taylor et al., 2002). Collectively, the above relate to the performance of the physical FRM system as well as of the actors responding to flood risk.

In addition to the causal relations, normative relations exist between the rules and the physical FRM system and actors. A relation is normative if one component includes a rule which provides a structuring context for the other component (Ottens et al., 2006). Actors use cognitive rules to shape perceptions of the future, and hence make decisions on adaptation in the present. Formal and normative rules also influence the behaviour and decisions of actors, as these are embedded in regulatory structures and social/organisational networks. Like the actors, the physical FRM system is structured by rules. For example, acceptable standards will limit the frequency or risk of flooding to a quantified level.

Continuity

Continuity is provided by the linkages and alignments between the different components of the FRM system. These linkages are the result of the responses of actor groups which produce and reproduce them. Their responses create and maintain the structural and non-structural components. For example, flood risk infrastructure is built and maintained by flood management organisations; flood risk regulation is made by governments; and expectations emerge from the way in which different groups perceive flood risk. The responses of different actor groups are coordinated and aligned to each other. This inter-group coordination is represented in the concept of socio-technical regimes (Geels, 2004). Socio-technical regimes account for continuity of existing systems through different mechanisms, as explained by Geels and Kemp (2007). Existing systems are stabilised by organisational rules, procedures and cognitive routines (Geels, 2005) and also by regulations and standards (Unruh, 2000). In addition, actors and social networks represent organisational capital and institutionalised power, which contribute to the continuation of existing systems. This is because of interdependent relationships, mutual expectations, organisational commitments and vested interests of existing organisations (*ibid*). Finally, the irreversible investments underpinning and the economics of use of large-scale engineering structures may lead to problems such as 'lock-in' to their use, as non-abandonable or non-adaptable infrastructures for decades into the future (Brown et al., 2011). As an example: sewer conduits with an excess capacity (over the design capacity) to deal with the worst case climate change scenario might subsequently be found to be excessively oversized. This could lead to regret about unnecessary costs (e.g., Ofwat, 2008; Arkell et al, 2011). It is these various mechanisms and structures that contribute to incremental adjustments to existing systems by following particular directions, leading to trajectories that are often path dependent. Often, however, these trajectories will lead to greater resilience, because incremental adjustements can accumulate over time and result in performance improvements. But, in some cases, these can lead to reduced resilience and maladaptation, where such trajectories are counterproductive to the system performance.

Innovation

System innovations emerge in niches as an effect of learning processes and network building. Niches are networks wherein it is possible to deviate from the rules in the existing regime (Rip and Kemp, 1998). The rules in niches are less specified and clear-cut than in regimes; there are only general rules and broad visions. This means there are less structuring effects and there is more space for learning (i.e., going in different directions and trying out variety). These general rules and visions become more specified as more is learnt about the new innovation. Furthermore, the social networks in niches are smaller and more precarious than in regimes. The building of social networks and constituencies to support the new innovation is, thus, an important internal niche process. In summary, niches contribute to innovation, because they provide space for key processes, such as the articulation and refinement of visions, learning processes and the build-up of social networks (Geels, 2004). The innovations in niches are directed to the problems of the existing regime and may eventually be used in the regime or even replace it. This is not easy, because the existing regime is stable in many ways; as explained above.

**Step 3B: Specify thresholds of identity**

Step 3B identifies the specific variables and thresholds that reflect changes in identity. For the existing FRM system to be considered resilient, the variables that define its identity should be maintained under the drivers specified in Step 1. This implies the system can have the same identity while also undergoing and adjusting to change, but only for change up to a critical threshold. A key variable that defines the identity of the FRM system concerns the system performance, i.e., its capability in terms of flood risk. The critical identity threshold occurs where the system performance is outside the acceptable risk level, as defined by law or decided by the stakeholders. If as a result of climate change, the existing FRM system can no longer deliver an acceptable risk, then it may be considered as a different system: i.e., it changes its identity. The magnitude of climate change beyond which the system identity changes then becomes a fixed point of reference against which the degree of resilience of the FRM system can be quantified. It is, finally, of note that the specific quantitative thresholds used to define identity changes will be emergent themselves,[5] because the standards and expectations will change in the future (for example, when external drivers and internal processes change the risk of flooding). It is, therefore, not possible to determine the potential boundaries of resilience (as defined below) with any certainty into the future.

**Step 3C: Assess when an identity threshold may be exceeded**

This step uses simulation models to determine the potential for a change in identity (and its magnitude) under alternative climate change scenarios. The outcome of this step will provide an estimate of the degree of resilience of the FRM system to climate change. It has been argued in step 3B that the FRM system has the same identity when the flood risk is maintained at an acceptable level through time. This will depend on the state of the FRM system. The level of risk in a particular system state can be assessed with the help of hydrological/ hydraulic models. It is, therefore, possible to identify the Adaptation Tipping Points (ATPs) of the existing physical FRM system by assessing the specific boundary conditions (e.g., the magnitude of climate change) under which acceptable standards and/or societal expectations toward flood risk may be exceeded. Because the FRM system (as a whole) is dynamic, the ATPs have to be identified not only for the existing physical FRM system, but also for the evolved system: i.e., after incremental adjustments have been implemented by the actors. The outcome of this step will provide an estimate of the potential boundaries of resilience. These refer the points of reference (i.e. pressure/pressures exerted on the system) where the exist-

---

[5] Emergence is defined as "the arising of novel and coherent structures, patterns and properties during the process of self-organization in complex systems" (Goldstein, 1999).

ing system, with the current adaptive strategy, will no longer be able to meet the objectives (often translated into acceptable standards). The identity approach for assessing resilience, thus, recognises that, in order to continue to function as required in the face of change, incremental adjustments will logically be triggered in the existing system before ATPs occur. Potential boundaries of resilience may be reached because of several reasons: e.g., the current dynamic adaptive strategy will become too expensive, ineffective, or will no longer meet technical or environmental limits (Franssen et al., 2011). Finally, the degree of resilience can be determined by assessing the potential for a change in identity (and its magnitude) under the different scenarios of climate change. If the system's identity is likely to be changed under these scenarios, then the system may lack resilience in certain respects. If the system is likely to maintain its original identity across a broad range of scenarios, then it has a higher degree of resilience.

## 2.3. Application

Step 3 of the resilience assessment is illustrated using the example of FRM for the Island of Dordrecht, the Netherlands. Dordrecht is the oldest, and was once the most important city of Holland (the region consisting of the provinces of North Holland and South Holland).

### Flood risk management for the Island of Dordrecht

Surrounded by a series of rivers and canals, the city of Dordrecht is effectively located on an island (Fig. 2.3). The Island of Dordrecht lies in the transition zone between a tidal reach and a river regime reach, where the extreme water stages are influenced by both the river discharge and the sea level. The flow direction depends on the discharge of the Rhine and (to a lesser extent) the Meuse. During low tides, water flows toward the sea through the Maeslant barrier in the Nieuwe Waterweg, the Hartel barrier in the Hartelkanaal and the locks in the Haringvliet. The Nieuwe Waterweg and the Hartelkanaal are open outlets, which can be closed off. The discharge at the Haringvliet locks depends on the Rhine discharge at Lobith. The locks are shut when the river discharge is low ($<$1,200 m$^3$/s), and fully open at a river discharge of 10,000 m$^3$/s. The flow direction changes when the Rhine discharge at Lobith is larger than 4,000 m$^3$/s. From this point onwards the river discharge starts to dominate the incoming tide flow.

Much of the city of Dordrecht is located in a single 37.1 km long dike ring protecting an area of about 70 km$^2$. The dike comprises the system of primary flood defence structures. Protection standards for dike ring areas have been established by national law (VenW, 2010) as the average exceedance frequency of the design water level that the flood defences must withstand. The legal protection standard for the Island of Dordrecht has been set at 1/2,000 per year. This average exceedance frequency constitutes a "formal" identity threshold.

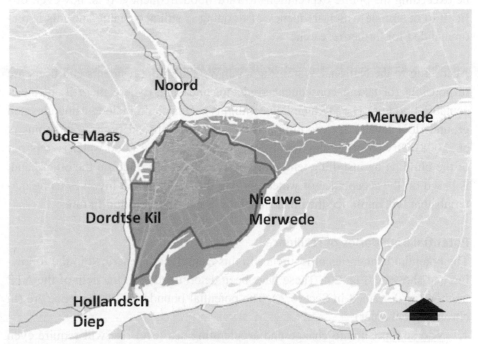

**Figure 2.3.** Rivers and canals surrounding the Island of Dordrecht

The statutory assessments of the primary flood defences are based on the protection standard and the corresponding design water levels. The findings of the Third Statutory Assessment for the Island of Dordrecht were that 28% of the flood defences are below standard (due to, among other, changes in hydraulic peak conditions) and require reinforcement (PZH, 2011). These reinforcement measures are part of, or will become part of, the Flood Protection Programme, which aims to strengthen inadequate flood defences for a 50-year period (IenM, 2011). The application of the 5-yearly statutory assessment of flood defences and the successful implementation of the Flood Protection Programme can thus be considered to be important mechanisms of continuity in FRM for the Island of Dordrecht. These mechanisms contribute to the maintenance of the system's identity in the face of change.

Part of the city of Dordrecht is situated outside the primary flood defences: the so-called unembanked areas. These areas are positioned at relatively high elevations in addition to being protected by the Maeslant barrier and Hartel barrier. The historic port area, with its quay heights between NAP (Amsterdam Ordnance Datum) +1.7 until +2.5 m, is the lowest-lying unembanked area. A "formal" identity threshold for the unembanked areas cannot be defined, because there are no legal protection standards for these areas. Socially, a critical identity threshold would be exceeding the public expectations toward flood frequency. It is, however, difficult to determine which frequency of flooding is still acceptable and this is recommended for further research.

According to the National Water Plan (VenW, 2009), the residents and users are responsible for taking consequence-reducing measures where there is an unacceptable flood risk (i.e., to maintain continuity). This could include using elevated ground floor levels, dry proofing and wet proofing ground floors (Gersonius et al., 2008). Dry proofing may involve shielding, where the flood water is kept out of the building by installing temporary barriers. Wet proofing, on the other hand, is based on the acceptance of water entering the building and involves using materials that will minimize the impact of flood water on fabric and fixtures.

## Potential boundaries of resilience

The potential boundaries of resilience in this example are described in the following and shown in Fig. 2.4. These have been determined with the help of the ATP method (as described in Sect. 6.2.). The potential boundaries of resilience are (in chronological order):

1. Voorstraat: Climate change and accelerating sea level rise will require even further reinforcement of the flood defences in the future. However, the costs for these reinforcement measures may be very high or the integration into the surrounding areas can be technically too complex and/or socially unacceptable. This is the case for the flood defence structure at the Voorstraat, which is a street in the historic centre of Dordrecht. At the Fourth Statutory Assessment (2016/2017), the Voorstraat will be rejected as a primary flood defence, because of its inadequate elevation. Strengthening the Voorstraat in a traditional way is likely to be socially unacceptable, as it would result in several years of construction in the heart of the historic city centre and a potential loss of historic character of the Voorstraat. These societal limits may lead to the occurrence of an ATP for the existing system, with the current adaptive strategy. Beyond this critical ATP, the system performance will be outside the acceptable risk level, reflecting some change of regulatory behaviour. This means that the FRM system may change its identity.

2. Historic Port Area: The frequency of occurrence of high-water situations in unembanked areas will increase with climate change and accelerating sea level rise. The historic port area of Dordrecht will be among the first unembanked areas to flood. Possible high-water levels in this area would mainly cause disruption and economic damage, but there is only a limited risk of injuries and casualties. As a first estimate, it is anticipated that the frequency of

flooding will remain acceptable until 2050 under the high climate change scenario (KMNI'06 W+ scenario), or until 2100 under the medium climate change scenario (KMNI'06 G scenario) (Van den Hurk, 2007). However, further research regarding the public expectations for flood frequency is needed to determine the timing of this ATP with more certainty.

3. <u>Maeslant barrier</u>: The Maeslant barrier plays an important role in FRM for the Island of Dordrecht. The barrier closes if the water level exceeds NAP +3.0 m at the outlet of the Nieuwe Waterweg or exceeds +2.9 m upstream at Dordrecht. The closing of the barrier ensures that the primary flood defences can continue to meet the legal protection standard of 1/2,000 per year. The Maeslant barrier has been designed to cope with 0.25 m sea level rise, and can be easily adjusted to cope with 0.5 m sea level rise. A sea level rise of 0.5 m will be reached around 2070 under the high climate change scenario, or around 2140 under the medium climate change scenario.

| At the earliest (W+) | 2010 | 2030 | 2050 | 2070 | 2090 |
| At the latest (G) | 2010 | 2060 | 2100 | 2140 | 2180 |

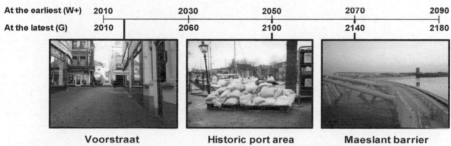

| Voorstraat | Historic port area | Maeslant barrier |

**Figure 2.4.** Timing of the critical ATPs for the current FRM strategy for the Island of Dordrecht

## 2.4. Discussion and conclusions

This chapter has demonstrated the utility of the identity approach for assessing resilience of STS, using the example of FRM for the Island of Dordrecht, the Netherlands. From these results, it has been shown that socio-technical resilience can be defined as the ability of the system to continue to function as required in the face of change. This definition implies that a STS is resilient when it can deliver performance over the assessment period without a change of identity by continuing compliance with standards and expectations. It is of note, however, that identity in this context is dynamic, as it responds to changes in standards and expectations.

Although the identity approach is useful for assessing resilience of STS, it also has limitations. An important limitation is that the selection of identity variables and thresholds is highly subjective and dependent on social values and interests (Cumming et al., 2005). Such normative decisions cannot be made by experts alone (Smith and Stirling, 2010), and should rather be the outcome of meaningful engagement with all the actors concerned, for example, via Learning Alliances (Ashley et al., 2012). Over time shifts in social values and interests can also alter perceptions of desired trajectories (Voß et al., 2007). This includes shifts in expectations of system performance. Because of this emergence, the threshold values used to define identity changes will also be emergent (i.e., identity is a dynamic property). It is, therefore, not possible to determine the potential boundaries of resilience with any certainty. A main research need is, thus, to understand how social values and interests evolve over time in response to socio-economic or climatic changes and to consider this in relation to the system's identity. Understanding these change processes will be crucial for managing resilience in relation to STS, not only those dealing with flooding.

# 3. Adaptive Policy Making

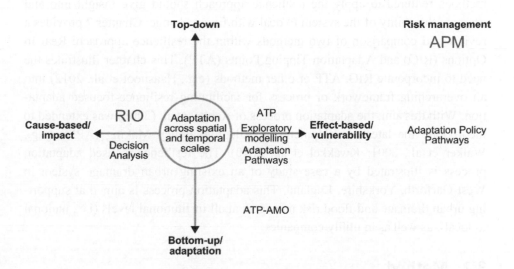

This chapter is based on:
Gersonius, B., Ashley, R. Pathirana, A. & Zevenbergen, C. 2012. Adaptation of flood risk infrastructure for climate resilience. *Proceedings of the ICE - Civil Engineering*, accepted.

## 3.1. Introduction

Over the 30-200 year lifetime of flood risk infrastructure there is significant sensitivity of its performance to climate change. Hence, when making decisions about infrastructure investment it will be important to account for climate change uncertainty and avoid closing off adaptation options that may be useful in the future. Flood risk infrastructure that is difficult to modify makes it potentially more costly to adapt in the future when more knowledge becomes available (HM Government, 2011). Utilising the resilience approach for climate impact and adaptation assessment increases the chances of avoiding such inflexible systems. The methods required to apply the resilience approach should give insight into and promote the ability of the system to deal with future change. Chapter 7 provides a review and comparison of two methods within the resilience approach: Real In Options (RIO) and Adaptation Tipping Points (ATP). This chapter illustrates the need to incorporate RIO, ATP or other methods (e.g., Haasnoot et al., 2012) into an overarching framework or process for facilitating resilience-focused adaptation. With this aim, the adaptation process of Ashley et al. (2008) was extended to incorporate the latest developments in Adaptive Policy Making (APM) (e.g., Walker et al., 2001; Kwakkel et al., 2010). The resilience-focused adaptation process is illustrated by a case study of an existing urban drainage system in West-Garforth, Yorkshire, England. This adaptation process is aimed at supporting urban drainage and flood risk managers at all institutional levels (i.e., national to local), as well as in utility companies.

## 3.2. Method

The adaptation process in Fig. 3.1 applies to all available methods within the resilience approach. It comprises 5 stages, which include the specification of a core and supporting strategy as well as the definition of a monitoring system to indicate whether the acceptable standard is likely to be compromised. The core strategy gives the responses and potential adaptations for providing infrastructure that delivers an acceptable risk through time; that is, for maintaining required performance. The supporting strategy addresses the internal processes (e.g., in available funding) and external changes (e.g., in regulations) that affect the performance of the core strategy. The monitoring system requires the identification of indicators (or: signposts (Walker et al., 2001)) and thresholds or trigger events for implementing incremental adjustments to the infrastructure or the way it is utilised or for reassessing the overall adaptive strategy.

The first stage is to develop an initial understanding and where possible quantification of the current flood risk, the acceptable risk level, and the capacity and options available to manage this, e.g. Brooks and Adger (2005). The acceptable risk level can be defined either according to the regulations (e.g., BSI, 1995-1998) or decided by the stakeholders involved. It is, furthermore, of note that the choice of the acceptable risk level is highly subjective and dependent on social values and interests. Therefore, such decisions should, where possible, be the outcome of meaningful engagement with all the actors concerned, for example, via Learning and Action Alliances (Newman et al., 2011).

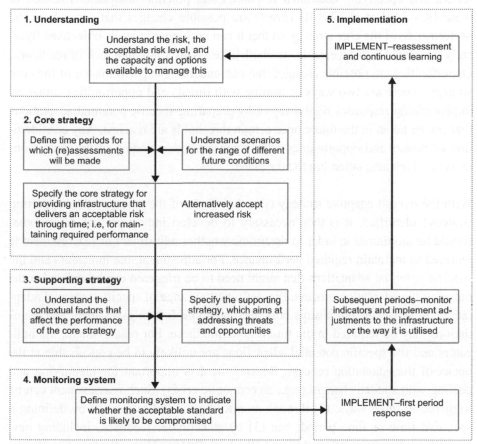

**Figure 3.1.** Adaptation process for FRM systems, showing the 5 stages (adapted from: Ashley et al., 2008; Kwakkel et al., 2010)

Following the understanding stage, the core strategy can be devised with the use of any of the available methods within the resilience approach (see Chapter 7). The use of these methods requires: the definition of the time periods for which risk and response (re)assessments will be made and the formulation of the climate change and socio-economic scenarios—as the resilience of the FRM system will crucially depend on these scenarios. The scenarios provide information to identify in advance the responses and potential adaptations for providing infrastructure that delivers an acceptable risk through time; i.e., for maintaining required performance.

In the next stage, the acceptance of and support for the core strategy is enhanced through the development of a supporting strategy. This stage is based on identifying in advance the threats and opportunities associated with the core strategy selected, and specifying additional responses and potential adaptations related to these (Kwakkel et al., 2010). Threats are possible changes that can reduce the performance of the core strategy so that it can no longer meet its objectives (typically translated into acceptable standards); i.e., with a possible lack of resilience. Opportunities are possible changes that can enhance the performance of the core strategy. There are two ways of dealing with threats and opportunities, either by implementing responses right away or by preparing specific potential adaptations that can be taken in the future once certain thresholds are reached. Any consideration of threats and opportunities relies upon an understanding of the wider contexts in which adaptation has to take place.

With the overall adaptive strategy (i.e., the whole of the core and the supporting strategy) identified, it is then necessary to develop indicators that define what should be monitored in order to determine whether adjustment or reassessment is required to maintain required performance. For adjustment, the indicators and the specific potential adaptations that might need to be triggered can be defined at the outset of the adaptation process. Pre-defining the range of specific potential adaptations is, however, not always possible, and some further assessment and negotiations may be required in the future to define these. For reassessment, the indicators and the specific potential adaptations are unlikely to be conceivable at the outset of the adaptation process. Reassessment is important for identifying and dealing with sudden changes (e.g., an economic crisis). Such reassessment can be triggered in three ways: (1) through monitoring via indicators; (2) by defining a specified time or time period; and (3) by stakeholder feedback, including new scientific information (Swanson and Bhadwal, 2010). The indicators for triggering reassessment are the same as those used for adjustment; however, the reassessment is often initiated as a result of the further negotiations that may be necessary in the future to define the specific potential adaptations. Time and stakeholder triggered reassessment are critical for identifying emerging problems even when the core strategy is performing well. Such trigger events could include the provision of new scenarios of e.g. climate change or the availability of new research concerning e.g. the effectiveness of responses. Finally, it is important to define in advance who should be responsible for the monitoring of performance and for comparing this with evolving acceptable standards as these will likely change. It is then also necessary to set out the options for responsibilities for adapting the system in the future and for conducting the reassessment. While this may also change in the future, at least the initial responsibilities should be defined.

Where there is a wish and the capacity to act among the actors concerned, the first-period responses are implemented right away, the monitoring system is established and the indicator information collection is initiated. After the implementation of the first-period responses, the adaptation process is suspended until a trigger event is reached (e.g., Environment Agency, 2009). As long as the original objectives and acceptable standards remain relevant, the responses to a trigger event are adjustments to maintain required performance or to help address threats and opportunities. In some instances, incremental adjustment might not be sufficient to continue with the core strategy (e.g., when the acceptable standard is likely to be compromised). For example, there may be changes in the objectives or significant unforeseen responses by other stakeholders. In such instances, the overall adaptive strategy should be reassessed and substantially changed or even abandoned. If reassessment is necessary, however, the learning from previous experiences should be incorporated into the adaptation process. A parallel component: active learning can help develop the capacity of different actors to accept different views on performance and to use different types of response and at different times in the adaptation process. Active learning is essential in order to develop effective and efficient responses to adapt to future change and this learning needs to be managed collectively across all stakeholders, ideally in partnerships such as Learning and Action Alliances (Ashley et al., 2012).

## 3.3. Application

The resilience-focused adaptation process is illustrated here by a case study of an existing urban drainage system in West Garforth, Yorkshire, England; the subject of a Defra Integrated Urban Drainage (IUD) pilot study (LCC et al., 2008). The West Garforth area is mainly low density residential and suffers repeated flooding. The area is drained by surface water sewers and highway drains, connected to three culverted watercourses (Fig. 3.2).

**Stage 1: Understanding**
The acceptable risk level for the urban drainage system has been decided by the stakeholders, Leeds City Council, Yorkshire Water Services and the Environment Agency, as: no flooding for a design storm with a 1 in 30 year recurrence interval. The system performance during the design storm was simulated using a hydrological/hydraulic model, SWMM (Rossman, 2004), verified for the Defra IUD pilot study. The simulation results predicted significant flooding at three main

locations: Recreation Ground/Barley Hill Road "A", Lowther Road "B", and Oak Drive/Station Fields "C", see Fig. 3.2. These results imply that the existing system already lacks resilience—and this will be further aggravated by climate change. Available adaptation options for reducing the flood risk to an acceptable level include: enlarging sewer conduits; building and upsizing storage facilities; and/or disconnecting back roof drainage in the sub catchments (e.g., Thorne et al., 2007). These options have been identified from a combination of public suggestion and professional modelling/recommendations.

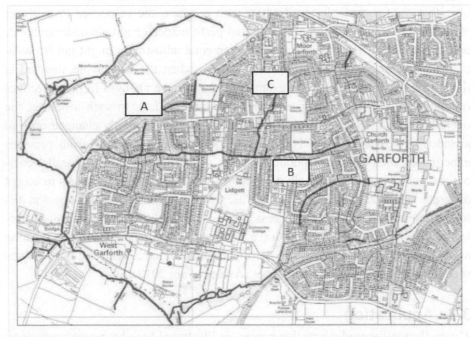

**Figure 3.2.** The West Garforth drainage network: Public Surface Water Sewer (in green), Highway Drain (in yellow), Open channel watercourse (in blue), Private culverted watercourse (in pink)

## Stage 2: Core strategy

The core strategy, consisting of a set of static adaptive strategies with the same initial configuration and different evolutionary (i.e., subsequent) configurations, was developed using RIO analysis; as described in detail in Sect. 4.4. The analysis considered that the system configurations can be built over 3 separate time periods, each of 30 years. This time period length has been selected because it typically takes a few decades before a climate change 'signal' can be detected in observed climate data. The core strategy is to build configuration A1 in the first time period. This includes enlarging sewer conduits (including the removal of adverse gradients) and building 6 new storage areas. Configuration A2 is built if the intensity of the design storm has gone up by 13% in either the second or third time period. This involves building 2 new storage areas, expanding 4 already-built storage areas and disconnecting stormwater from 213 back roofs. Configuration A3 is built only when the rainfall intensity has increased by 28% in the third time period. This includes building 1 new storage area, expanding 3 already-built storage areas and disconnecting stormwater from a further 137 back roofs.

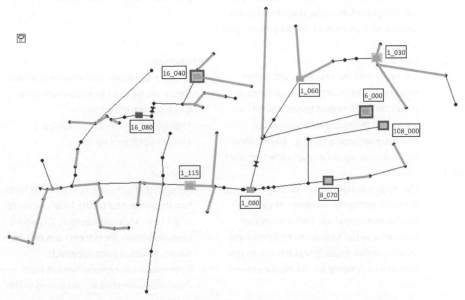

**Figure 3.3.** Schematic of the urban drainage system, including system configurations A1 (in green), A2 (in orange) and A3 (in red). The lines represent sewer conduits; the dots represent manholes; and the boxes represent storage facilities

## Stage 3: Supporting strategy

The contextual factors that affect the performance of the core strategy include:
1) The flood management responsibilities are defined in a clear and holistic way
2) The actors are willing and able to act
3) The actors are able to secure the funding necessary for the adaptation options
4) The adaptation options identified remain available in the future and perform as required when implemented
5) The actors remain committed to the active learning process (even though the individual/institutional actors may be different over time)

The core strategy has many different threats and opportunities, and these can be directly related to the contextual factors affecting the performance of the core strategy. Table 3.1 presents some of the main threats and opportunities of the core strategy, as well as possible responses and potential adaptations to these.

**Table 3.1.** Examples of threats and opportunities and possible responses and potential adaptations to these

|  | Threats and Opportunities | Responses and potential adaptations to these |
|---|---|---|
| Ad 1: | Threat:<br>Flooding responsibilities are split between many different individuals and organisations (see Fig. 2). The responsibilities are not integrated. No clear responsibility for dealing with sources of flooding from open land. | Response:<br>- Develop a national, clear and holistic definition of flooding responsibilities; preferably enshrined in statute. |
| Ad 2: | Threat:<br>It is likely that the residents will be the most reluctant to act. This is particularly significant with regard to upstream residents and their potential to alleviate downstream problems (e.g., by disconnecting roof drainage from the culvert system). | Responses:<br>- Adjust regulations and standards where these can be seen to encourage or require uptake of options.<br>- Engage the residents to develop a greater uptake of options. |
| Ad 3: | Threat:<br>The main mechanism for funding is through the local authority. However, they have insufficient resources. Yorkshire Water Services and the Environment Agency are unable, within existing regulations, to provide any co-funding for the options identified. | Responses:<br>- Fund potential future upgrades via local tax, implemented by the local authority or EA (e.g., localism agenda). Or, where responsibilities for systems are split between actors, a joint approach.<br>- Reconsider the enmainment of high flood-risk culverted watercourses in the area. |
| Ad 4: | Threat:<br>Land might not be available when needed, as a result of development or of opposition against options that impact on people's lives (e.g., through a swale in the street). | Responses:<br>- Reserve unused areas (e.g., grassed) for future surface storage, possibly with some temporary function linked to e.g. green infrastructure benefits.<br>Potential adaptations:<br>- Engage vigorously with the residents to make sure that the options are effectively 'owned' locally. |
| Ad 4: | Opportunity:<br>Some implemented options may perform better than expected. | Potential adaptations:<br>- Delay potential adaptations that are found to be no longer necessary. |
| Ad 5: | Threat:<br>Some of the actors may lose interest in maintaining sufficient learning at the right time and over a long enough time period to be able to make the correct decisions when and where required to keep up with the deteriorating performance of the system. | Responses:<br>- Where the risk associated with *not* maintaining required levels of performance is likely to be significant, consider the establishment of a long-term learning alliance, incorporating all actors (Ashley et al., 2011). |

## Stage 4: Monitoring system

Table 3.2 shows examples from the proposed monitoring system, and the adjustment and reassessment necessary when a certain threshold or trigger event is reached. In addition to the monitoring via indicators, it is important to conduct reassessment at a frequency of about 5 years to identify any emerging problems.

**Table 3.2.** Examples of the monitoring system

| Monitoring system | Adjustment and reassessment necessary |
| --- | --- |
| Monitor the developments in information about the changes in intensity of the design storm | If the intensity has gone up by 13%, build configuration A2. If the intensity has increased by 28%, build configuration A3. If the intensity has increased by more than 28%, reassess the overall adaptive strategy. |
| Monitor effectiveness of installed rain cisterns. | If rain cisterns appear to be inadequate in the future, reassess the strategy as regards their usage. |
| Monitor the public acceptability and engagement with options implemented and those that may be used in the future. | If the acceptability of certain options is declining, engage more vigorously with the residents.. |
| Monitor the willingness and actual progress of actors participating in the learning process. | If the key actors are not willing or able to participate (e.g., due to loss of staff or capacity), reassess the overall adaptive strategy. |

## Stage 5: Implementation

As part of the implementation stage, the first-period responses are implemented right away. System configuration A1 is constructed and the responses related to the threats are implemented (e.g., collecting the local tax). In addition, the monitoring system is established and the indicator information collection is initiated. The construction of the evolutionary system configurations and the implementation of the specific potential adaptations related to the threats and opportunities should begin only if triggered by climate change or (other) external developments (e.g., engaging more vigorously with the residents about options). In some instances, if e.g. incremental adjustment has not been adequate to continue to maintain required performance, the reassessment of the overall adaptive strategy might be required. If reassessment were required, the decision makers would reiterate through the stages of the adaptation process to develop a new overall adaptive strategy.

## 3.4. Discussion and conclusions

Urban drainage and flood risk managers usually deal with uncertainties about current climate by defining acceptable standards, either based on likelihood or based on a broader risk-based approach (i.e., taking account of the likelihood as well as consequences). However, climate change introduces additional uncertainty. This is no longer simply something to be reduced through more scientific research, although this can help. A range of methods to account for climate change uncertainty are being further developed, but the majority of these are not yet fully operational. As an example: RIO analysis, as used in this chapter to develop the core strategy, does not yet provide a sufficiently developed adaptation process that can be readily implemented in practice—despite recommending its' use in England and Wales (HM Treasury and Defra, 2009) and further work is underway to address this (EA, 2012). This demonstrates that more attention has to be paid to putting these methods more effectively into practice. This chapter has presented a 5-stage adaptation process, termed APM, that may be applied using any of the methods available within the resilience approach and has illustrated its use in an urban drainage case study; although APM could be applied to decision making involving any technical/infrastructure system. The experience of applying this method to the case study was largely positive, because of its comprehensiveness (e.g., having well-defined stages and steps) and inclusiveness (e.g., addressing both threats and opportunities). Nonetheless, the implementation of the resulting core strategy has been less successful in this case, mainly due to a lack of funding. In West Garforth, the main funding mechanism (other than from the residents) is through the local authority, which has insufficient resources, although it has funded some renovations to the culvert system. Yorkshire Water Services and the Environment Agency have been unable, within existing regulations, to provide any co-funding for the adaptation options identified and this has been identified as a major threat of the core strategy (refer to Table 3.1), although under the Flood and Water Management Act 2010 and Defra implementation guidelines, cross-stakeholder funding of such schemes is becoming the norm (e.g. Defra/EA, 2011). A potential response (from the supporting strategy) to this threat is to (re-)consider the enmainment of high flood-risk culverted watercourses, which will open up the possibility of the Environment Agency co-funding some of the options—possibly with the support of local tax funds or local levy funding. If this response will also prove to be unsuccessful, then the overall adaptive strategy has to be reassessed—and this should take account of any lessons learnt. It could, for example, be found that the acceptable risk level for the urban drainage system has been set too stringently with respect to the available resources. In the future, the acceptable standard may, therefore, be allowed to decline: i.e., because of the learning experience. In this regard, it is also recommended to consider (in any future study) the headroom that is available in terms of the residual performance of the entire system (minor/major) over and above the capacity of the (engineered) urban drainage system. The minor/major system performance can be expected to be greater than (just) the capacity of the (engineered) urban drainage system, because exceedance flows are contained within surface pathways (e.g. road surfaces). As such, the integrated management of urban drainage and surface water flooding can help to save costs of adaptation, while at the same time enhancing resilience to climate change.

# 4. Real In Options

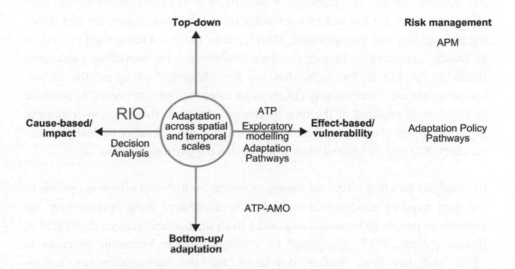

This chapter is based on:
Gersonius, B., Ashley, R., Pathirana, A. & Zevenbergen, C. 2012. Climate change uncertainty: building flexibility into water and flood risk infrastructure. *Climatic Change*, accepted.

## 4.1. Introduction

The planning/modification of flood risk infrastructure is costly and usually takes some time to implement, so the consequences of the choices made as to the form and function of these systems have to be lived with over many decades. In the past this did not matter as the rate of change of external drivers was slow in comparison with the lifetime of the systems and these continued to function as required over the long term. Nowadays, the uncertainties associated with the process of making decisions for infrastructure investment can be significant and arise from, amongst other factors, a lack of knowledge about primary external drivers, like climate change. The planning or modification of such systems should thus effectively allow for the lack of knowledge by explicitly designing for and building in flexibility into infrastructure. 'Real Options' (RO) is a recognised procedure to handle uncertainties in infrastructure investments by providing managerial flexibility. A RO is 'the right—but not the obligation' to adjust the technical/infrastructure system in ways likely to be more resilient, as needed to continue to function as required in the face of change. As such, these options represent physical choices about the system that can provide the flexibility to deal with uncertainty. RO may be utilised even when modifying existing systems.

RO analysis provides a rational means to decide on the most effective options to maintain required performance and when to implement these options over the assessment period. RO analysis originated from the options analysis developed in finance (Myers, 1984), recognised by a Nobel Prize in Economic Sciences in 1997, and has been further developed for the management of technical/infrastructure systems since the 1990s (Dixit and Pindyck, 1994) (Trigeorgis, 1996) (Amram and Kulatilaka, 1999) (Antikarov and Copeland, 2001) (Agusdinate, 2008). The analogy between financial market interventions and the decisions required for major infrastructure investments, allows the models from financial options analysis such as the Black-Scholes model (Black and Scholes, 1973) and the Cox, Ross and Rubinstein binomial model (Cox et al., 2002), to be used to inform how best to manage technical/infrastructure systems in the presence of uncertainty. RO analysis has more recently been applied to the (re)design of technical/infrastructure systems (Zhao and Tseng, 2003) (Zhao et al., 2004) (Wang and de Neufville, 2004) (Buurman et al., 2009) (Gersonius et al., 2010). However, the focus of the RO analysis applied to the (re)design of technical/infrastructure systems is different from that of traditional RO analysis. Traditional applications analyse options from a financial perspective treating the system configuration as a black box, such as deciding whether to make an investment now or in the future. Whereas when applied to the (re)design of technical/infrastructure systems, RO analysis focuses on making changes to the system configuration in response to reductions in uncertainty through future learning. This is known as Real In Options (RIO) analysis (De Neufville, 2003). In this type of application, the options relate to the technical characteristics of the system, which cannot be known if the system configuration is treated as a black box. The development of RIO analysis provides a framework to find out which flexibilities, that permit the system to be adapted over time, are worth their cost. Previously this had not been possible—with the consequence that flexible design was traditionally neglected (de Neufville, 2004). This chapter has applied RIO analysis to the selection and timing of options for the modification of an urban drainage system as an example of how best to adapt these systems to provide greater climate change resilience.

## 4.2. Climate adaptation for flood risk

Conventional planning/modification of flood risk management (FRM) systems is typically based on the use of probability density functions (PDFs), which describe the expected annual frequency of a rainfall or river flow event under known conditions. PDF estimation assumes that the rainfall or river flow observed over a long time period is at least partially stationary in a statistical sense. Under the quasi-stationary hydrology assumption, observed (historic) time series of hydro-climatic variables can be used to characterise events in the future. However, an increasing lack of stationarity in hydro-climatic phenomena, and hence in predictability of future pressures and impacts, makes it necessary to modify the conventional planning/modification of FRM systems. In this regard, Milly et al. (2008) have argued that there is a need to use non-stationary probabilistic data for hydro-climatic variables in optimisation procedures for these systems. Few studies so far explicitly make use of large-ensemble climate modelling data in assessing and managing future impacts. One study that does take a probabilistic approach is that of New et al. (2007). They have applied a full end-to-end uncertainty analysis in climate impact and adaptation assessment. This approach propagates the uncertainty in climate models through to hydrological models to assess impacts. Adaptive strategies are then developed based on the likely impacts of climate change on the system of interest. Brugnach et al. (2008) argue, however, that uncertainty propagation is inadequate whenever the system to be managed is a complex adaptive system. Such a system is able to learn and adapt to a changing context. The flaw of uncertainty propagation is that it assumes that the system being managed continues unchanged throughout the assessment period. This assumption fails to recognize that the learning and adaptation mechanism reduces future uncertainties associated with the system.

The uncertainty propagation method is still adequate, however, for the development of a static robust strategy to climate change, that once implemented is not actively managed. The static robust strategy is sometimes applied for the implementation of adaptation comprising large-scale hard structural measures with high (fixed) capital cost, such as large embankments, major sewers, or similar potentially irreversible measures. The selection of a static robust strategy usually requires the technical/infrastructure system to be initially designed to accommodate the worst case climate change scenario. This implies the adoption of a 'headroom' methodology (Ingham et al., 2006). Headroom is the excess capacity added on to the 'design capacity' to allow for future uncertainties that cannot be resolved at the present time and is standard engineering practice; frequently known as a 'factor of safety'. Introducing this headroom capacity into the technical/infrastructure system will help ensure that the required levels of performance can be achieved even with uncertainty. The static robust strategy is thus designed to function without any performance monitoring and significant readjustment of management throughout the system lifetime.

Currently there is not an agreed procedure for the development of a dynamic adaptive strategy to climate change. This strategy consists of a set of static adaptive strategies with the same initial configuration and different evolutionary configurations. A sequence in which a dynamic adaptive strategy can move from one configuration to an alternative configuration is referred to as an adaptation pathway (Haasnoot et al., 2012). The dynamic adaptive strategy confers the ability, derived from e.g. built-in flexibility (i.e., keeping options open), to adjust to future uncertainties as they unfold. This reduces the effect of decisions made at the start of the adaptation process that might subsequently be found to be not the best, resulting in e.g. unnecessary costs of potentially irreversible measures. However, building flexibility into infrastructure generally brings an increase of costs. The additional costs of flexibility are associated with the options that must be designed and build in with extra cost. Hence, it is required to assess which flexibilities, that permit the engineering system (re)design to be adjusted over time, are worth their cost. This means that the flexibility value has to be estimated. As RIO analysis explicitly accounts for the flexibility built into alternative adaptive strategies, the use of this procedure is proposed here for the development of a dynamic adaptive strategy to climate change. This is the novel contribution of this chapter.

## 4.3. Method

RIO analysis (Wang and De Neufville, 2004) determines the flexibility value based on a probability distribution of the uncertainties in future time periods. It recognises that uncertainty cannot be completely resolved over time, but rather that due to advances in knowledge the probability distribution will be adjusted in the future. In this sense, learning about climate change plays a significant role in informing the size and timing of investments. For RIO analysis, a number of characteristics are required that the traditional RO theory does not deal with, including the technical details of and interdependency/path-dependency amongst options. Path-dependency is that the value of an option depends on whether some other part of the technical/infrastructure system was or was not built. This feature implies that the recombining tree for financial options is insufficient for RIO analysis. In the recombining binomial tree (right tree in Fig. 4.1), the valuation of an option on each node of the tree is path-independent. This means that the valuation of the option on a certain node is the same for any path leading into that node. Wang and De Neufville (2004) proposed breaking the recombination structure of the tree to deal with path-dependency features. In a non-recombining binomial tree (left tree in Fig. 4.1), each path is depicted separately.

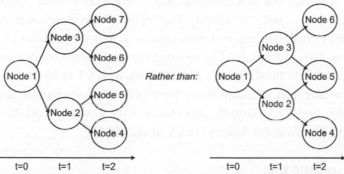

**Figure 4.1.** Binomial tree for RIO analysis (left) and financial options analysis (right)

The set of procedures to develop and apply a RIO optimisation model has been developed by Wang and De Neufville (2004), and is modified here for the adaptation of FRM systems to climate change. It comprises of the following steps:

1. Specify the scenario tree for the stochastic process that the uncertain variable follows. In the case example of the modification of an urban drainage system (see Section 4.4), the uncertain variable is the rainfall intensity of the design storm. It is assumed that the change in rainfall intensity follows a Geometric Brownian Motion (GBM) in order to develop the scenario tree. The GBM assumption is explained / justified in the next section.

2. Identify the potential options, or flexibility, in the technical/infrastructure system. The flexibility is provided by the design variables that can be changed after the initial implementation of a configuration. In the case example, the RIOs arise from the possibility of replacing sewer conduits, building and upsizing storage facilities, and/or disconnecting back roof drainage in the

sub catchments (Evans et al., 2004; Thorne et al., 2007). Therefore, these options give the opportunity to adjust the urban drainage system to future uncertainties as these unfold.

3. Formulate the RIO optimisation problem in terms of its objectives, constraints and decision variables. Here, the objective function is to minimise the expected life cycle cost for the dynamic adaptive strategy. The constraints in the optimisation process are the capacity, technical and RIO constraints. These are discussed in the next section.

4. Establish and run the RIO optimisation model.

## 4.4. Application

The context for the case study application is taken from an existing urban drainage system in West Garforth, Yorkshire, England. Recent guidance for climate adaptation in England and Wales recommends the application of RO analysis but does not provide any specific details as to how this should be applied (Defra, 2009). The analysis is carried out in a computer model that uses Genetic Algorithms, NSGAII (Deb et al., 2002), to identify the "optimal" dynamic adaptive strategy that continues to maintain required performance across the range of climate change scenarios. This is determined by meeting the acceptable standard for FRM, in this case no manhole flooding for a design storm with a 1 in 30 year recurrence interval. This acceptable standard was selected after discussions amongst the stakeholders within the West Garforth area (Leeds City Council, Yorkshire Water Services and the Environment Agency) (LCC et al., 2008).

**Step 1: Specify the scenario tree**
Due to climate change, there is uncertainty about the rainfall intensity of the design storm for future time periods. To represent this uncertainty it has been assumed that the change in rainfall intensity follows a GBM, which is a continuous-time stochastic process where the logarithm of the uncertain variable follows a random walk (i.e., Brownian motion). This assumption has the property that the variance of the uncertain variable increases over time. GBM is one of the most important basic notions of stochastic processes, and in particular, is the basis of options theory. A trinomial tree is used to simplify the stochastic process by means of a time-discrete representation of the change in rainfall intensity over a small number of time periods. The trinomial tree states that the rainfall intensity, denoted by $I$, can only move up, mid, or down during a given time period leading to a new value, $I_u$, $I_m$ or $I_d$. There is a probability of the up movement, mid movement, and down movement. Here, a particular series of up, mid and down movements is called a climate change path. Since the up, mid and down movements are independent events in terms of probability, the calculation of the prob-

ability associated with a climate change path is relatively simple. To be consistent with the stochastic process, the up factor $u$, mid factor $m$, and down factor $d$, together with the move-up probability $p_u$, move-mid probability $p_m$, and move-down probability $p_d$ need to be set in the following manner:

$$u = e^{\sigma\sqrt{2t}}, \quad m = 1, \quad d = e^{-\sigma\sqrt{2t}}$$

$$I_u = I \cdot u, \quad I_m = I, \quad I_d = I \cdot d$$

$$p_u = \left( \frac{e^{\mu t/2} - e^{-\sigma\sqrt{t/2}}}{e^{\sigma\sqrt{t/2}} - e^{-\sigma\sqrt{t/2}}} \right)^2$$

$$p_d = \left( \frac{e^{\sigma\sqrt{t/2}} - e^{\mu t/2}}{e^{\sigma\sqrt{t/2}} - e^{-\sigma\sqrt{t/2}}} \right)^2$$

$$p_m = 1 - p_u - p_d$$

Where $\sigma$ is the volatility, which scales the uncertainty in future rainfall intensity, $t$ is the length of the time period, and $\mu$ is the drift rate, which is the average magnitude of climate change per time unit. The drift rate $\mu$ and volatility $\sigma$ for the change in rainfall intensity have estimated by using climate modelling data that is available from the UKCP09 probabilistic projections (Murphy et al., 2009). This was done by producing 30-year hourly time series (100 in total) for both the 1990s (i.e., the baseline period) and 2080s with the UKCP09 Weather Generator, and then analysing these time series with the annual maximum method (Chow et al., 2005) to obtain the rainfall intensity of the design storm for both periods. Whilst UKCP09 provides probabilistic data on the hydro-climatic variables, this data is conditional on the high, medium and low climate change scenarios. As no information is available on the likelihood associated with the scenarios, it has been assumed that all scenarios are equally likely (the Laplace criterion). The cumulative distribution function of the change in rainfall intensity for the period 1990s-2080s is shown in Fig. 4.2. The change in rainfall intensity over the 90 year period appears to fluctuate randomly around a mean value, indicating that the climate change rate may follow a GBM process. Application of the Shapiro-Wilk W test (Shapiro and Wilk, 1965) confirms that the null hypothesis of a normal distribution cannot be rejected at the 0.05 level of significance (see also Fig. 4.3).

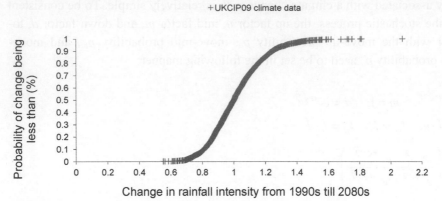

**Figure 4.2.** Cumulative distribution function of change in rainfall intensity

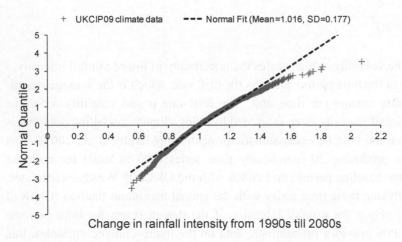

**Figure 4.3.** Normal probability plot of the change in rainfall intensity, used for the Shapiro-Wilk W test

Based on the mean and standard deviation of the normal approximation in Fig. 4.3, the drift rate $\mu$ and the volatility $\sigma$ have been calculated as follows:

$$\mu = \frac{(1.016 - 1)}{90} = 0.018\% \ per \ year$$

$$\sqrt{90}\sigma = \frac{\ln\left(\frac{1.016 + (2*0.177)}{1.016}\right)}{2} \qquad or \qquad \sigma = 1.573\% \ per \ year$$

Three time periods of 30 years have been selected to develop the trinomial tree. The reason for selecting a 30-year time period is that it typically takes a few decades before a climate change 'signal' can be detected in the observed climate data—mainly due to the existence of the multi-decadal climate variability (Hawkins and Sutton, 2009). The resulting trinomial tree representation is shown in Table 4.1.

**Table 4.1.** Scenario tree representation, including the "optimal" dynamic adaptive strategy. The strategy is to build configuration A1 initially, to build configuration A2 if the rainfall intensity has gone up by 13%, and to build configuration A3 only if the rainfall intensity has increased by 28%

| Start of time period 1 (1990s-2020s) | | | Start of time period 2 (2020s-2050s) | | | Start of time period 3 (2050s-2080s) | | |
|---|---|---|---|---|---|---|---|---|
| Change in rainfall intensity | % of each path | Config built | Change in rainfall intensity | % of each path | Config built | Change in rainfall intensity | % of each path | Config built |
| | | | | | | 1.28 | 6.6% | [A3] |
| | | | 1.13 | 25.7% | [A2] | 1.13 | 12.9% | Not any |
| | | | | | | 1.00 | 6.2% | Not any |
| | | | | | | 1.13 | 12.9% | [A2] |
| 1.00 | 100% | [A1] | 1.00 | 50.0% | Not any | 1.00 | 25.0% | Not any |
| | | | | | | 0.89 | 12.1% | Not any |
| | | | | | | 1.00 | 6.2% | Not any |
| | | | 0.88 | 24.3% | Not any | 0.89 | 12.1% | Not any |
| | | | | | | 0.78 | 5.9% | Not any |

## Step 2: Identify the potential options

The urban drainage system of West Garforth consists of 85 sewer conduits, 9 possible storage facilities and 15 sub catchments (Fig. 4.4). Associated with these structural components are the design variables, which define the possible system configurations. The potential RIOs in the urban drainage system have been identified by specifying the design variables that can be changed after the initial implementation of a configuration, as well as their range of flexibility. It was considered that RIOs arise from the possibility of replacing sewer conduits, building and upsizing storage facilities, and/or disconnecting back roof drainage in the sub catchments (Table 4.2). Therefore, these options give the opportunity to adjust the urban drainage system to future uncertainties as they unfold. It is of note that when a static robust strategy is adopted, then the design variables have to be considered fixed over the assessment period. That is, the urban drainage system cannot be adjusted over time.

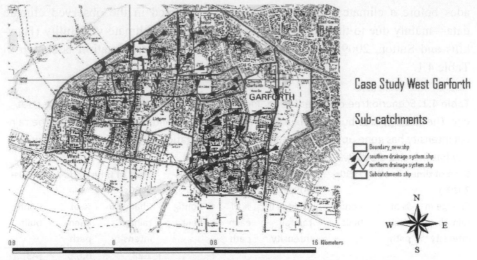

Case Study West Garforth

Sub-catchments

Figure 4.4. Scheme of the drainage network

Table 4.2. Design variables for the urban drainage system. Each of the variables is only allowed to vary between practical upper and lower bounds. This is represented by the 'range' in Table 4.2. The 'step' in Table 4.2 corresponds to the capacity increment that is (arbitrarily) used in this work. For physical and technical reasons, the design variables can be increased but not decreased

| Component | Design variable | Unit | Range | Step |
|-----------|----------------|------|-------|------|
| Sewer conduits | Conduit diameter | [m] | 0.15 - 0.90 | n/a |
| Storage S1_080 | Storage area | [m2] | 0 - 250 | 250 |
| Storage S1_060 | Storage area | [m2] | 0 - 250 | 250 |
| Storage S1_115 | Storage area | [m2] | 0 - 1000 | 250 |
| Storage S6_000 | Storage area | [m2] | 0 - 2000 | 500 |
| Storage S108_000 | Storage area | [m2] | 0 - 2000 | 500 |
| Storage S16_040 | Storage area | [m2] | 0 - 2000 | 500 |
| Storage S16_080 | Storage area | [m2] | 0 - 1000 | 250 |
| Storage S8_070 | Storage area | [m2] | 0 - 2000 | 500 |
| Storage S1_030 | Storage area | [m2] | 0 - 1000 | 250 |
| Sub catchments | Disconnected roof area | [%] | 0 - 60 | 15 |

## Step 3: Formulate the RIO optimisation problem

There exist different design solutions to build in flexibility into the urban drainage system. This implies that cost-effective engineering system (re)design is complex due to the large number of design variables and their interactions, coupled with the considerations of capacity, technical and RIO constraints. Here, the use of evolutionary optimisation techniques offers the potential to identify the "optimal" dynamic adaptive strategy. Genetic Algorithms can be used to select the set of RIOs and the rules for their implementation as part of an integrated simultaneous procedure (Buurman et al., 2009).

The objective to optimise is the expected life cycle cost for the dynamic adaptive strategy. In the procedure for RIO analysis, the expected costs of the set of static adaptive strategies are averaged over all climate change paths based on the probabilities derived from the stochastic process in order to obtain only one life cycle cost for the dynamic adaptive strategy. The following cost components are considered in the optimisation process:

> Initial capital cost: This corresponds to the investments needed to implement the initial configuration.
> Evolution cost: This corresponds to the necessary investments to implement the next adaptation step(s) in the dynamic adaptive strategy.
> Damage cost: This relates to the damage costs from flood events and climate change.

The following cost functions have been defined based on standard approaches to calculate the initial capital costs and the evolution costs between the previous and evolutionary configuration from the assumptions made in the original Defra IUD study (LCC et al., 2008):

£ per m conduit replacement = $504 * D + 135$ (where $D$ is the conduit diameter in m)
£ per $m^3$ (additional) storage capacity = 70
£ per (additional) roof disconnected to a rain barrel = 210

The evolution costs are discounted on the basis of a discount rate of 3.5%, based upon HM Treasury's discount rate (Treasury, 2003). Furthermore, the damage costs associated with an infeasible static adaptive strategy (i.e., a strategy that does not meet the acceptable standard for FRM) are set at a very large value to ensure that this strategy does not get selected in the optimisation process.

There are a number of capacity, technical and RO constraints in the optimisation process. The capacity constraint makes sure that the capacity in place in each time period and each climate state is always sufficient to meet the acceptable standard in that state. This constraint includes the need to have a climate resilient urban drainage system. Hydraulic simulation is used to test the candidate configurations with design variables in terms of available capacity of the drainage network. For this purpose, the RIO optimisation model (refer to step 4) is integrated with a hydrological/hydraulic model, SWMM (Rossman, 2004). SWMM has been selected due to the public accessibility of its source code and its well-accepted modelling capability and accuracy. The technical constraint is comprised of a construction constraint on the allowed values for the design variables of the drainage network. The range for the most important design variables is shown in Table 4.2. Two

RIO constraints make sure that: (i) any RIO can only be implemented at most once in any climate change path; and (ii) the climate state in any time period can only be distinguished by information available up to that period. The latter constraint is known as the non-anticipativity constraint (Wang and de Neufville, 2005).

It should be noted that the procedure for RIO analysis handles path-dependent options, in which the option value depends not only on the value of the uncertain climate variable but also on the particular climate change path. This means, for example, that if the magnitude of climate change has been large in preceding time periods, the decision makers may have been forced to increase the capacity of the drainage network in order to meet the acceptable standard, as they might not have done if the magnitude of climate change had been small in those periods.

Although the climate state is observed at the end of each time period, investment decisions have to be taken before the state is known due to the large lead-in time of implementing flood risk infrastructure. In this regard, headroom is used to ensure that the required level of performance is met whatever the climate state in the subsequent time period may be. Headroom refers to the excess capacity added on to the design capacity of the system to allow for future uncertainties that cannot be resolved at the present time (Ingham et al., 2006). This is translated into a headroom allowance equal to the up factor.

In summary, the complete formulation of the optimisation problem is as follows:

Objective:
> Minimise expected life cycle cost (= initial capital cost + evolution cost) under climate change uncertainty

Constraints:
> Capacity constraint to prevent flooding
> Technical constraints on the allowable values for the design variables
> Two RIO constraints

Decisions:
> The system configuration in initial time period
> The evolutionary system configurations in the subsequent time periods

Input values:
> Existing drainage network design
> Current rainfall intensity

> Future changes in rainfall intensity due to climate change
> Investment cost functions

### Step 4: Establish and run the RIO optimisation model

The RIO optimisation model has been implemented in a computer program written in C++ which was then used to automatically generate and eventually identify the "optimal" dynamic adaptive strategy. The genetic algorithm for the generation and analysis of the set of static adaptive strategies is based on NSGAII (Deb et al., 2002), and was set-up using a population of 100 individual solutions and running for 1,500 generations (mutation rate = 0.01 and crossover rate = 0.4). Mutation is the operation that randomly alters an individual solution from one generation to the next. A mutation rate of 0.01 means that any selected solution has a 1 % chance of being mutated during a generation. Crossover is the operation that combines at least two individual solutions in order to create new solutions for the next generation. A crossover rate of 0.4 means the 40 % chance of crossover occurrence.

## 4.5. Results

The outputs from the RIO optimisation model give the "optimal" dynamic adaptive strategy to climate change (Table 4.1). The saw-tooth effect in flood probability/ risk, as a consequence of taking a dynamic adaptive strategy, is represented in Fig. 4.5. The Expected Net Present Cost (ENPC) of implementing the dynamic adaptive strategy is about £1.70 million. The static robust strategy, which is associated with a fixed (re)design that accounts for the worst case path of climate change, has a deterministic Net Present Cost (NPC) of £2.13 million. These outputs show that if there is the possibility to incrementally adjust headroom allowances in the light of future learning, the (expected) costs of adapting to climate change can be reduced by more than 20% compared to when no learning and adaptation take place.

Two mechanisms explain this economic benefit. The first mechanism is that the dynamic adaptive strategy avoids potentially irreversible investments by building a more modest configuration initially with the option to expand. Another mechanism is that the costs for building the next adaptation step(s) in the dynamic adaptive strategy are discounted. In options theory, this benefit is called the flexibility value.

The dynamic adaptive strategy also minimises the maximum possible error with respect to maladaptation, which results from unnecessary costs as a consequence of irreversible and subsequently found to be not the best investments. The estimated maximum regret avoided with respect to maladaptation errors is equal to the NPC of the static robust strategy minus the initial capital cost of the dynamic adaptive strategy, which amounts to about £470.000 million (22% of the cost of the static robust strategy) in this case.

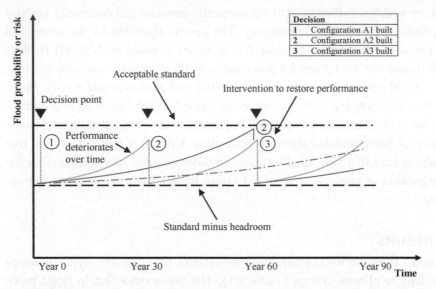

| Decision | |
|----------|-----------------------|
| 1 | Configuration A1 built |
| 2 | Configuration A2 built |
| 3 | Configuration A3 built |

**Figure 4.5.** Graph of flood probability/risk with time as a consequence of taking a dynamic adaptive strategy. This strategy will move from configuration to configuration as capacity increments are implemented. Key technical characteristics of configurations A1 to A3 are shown in appendix A

## 4.6. Discussion and conclusions

This chapter has introduced RIO analysis as a method that combines the resilience approach with the cause-based approach to identify an "optimal" set of static adaptive strategies in response to advances in knowledge about climate change. RIO analysis does not aim at optimising for some average or worst-case scenario, but actively adapts to changing conditions (Buurman et al., 2009). It has been demonstrated that RIO analysis has significant potential to provide economic benefits to FRM and that the possibility to build in flexibility into infrastructure should be taken into account in optimisation procedures. Being able to take advantage of these economic benefits is particularly relevant now, as scientists, policy makers and politicians are calling for the development of climate-proof solutions in some of the most important policy areas (e.g., EuropeanCommission, 2009). It is, furthermore, of note that the size of the benefits (i.e., the flexibility value) is expected to be sensitive to different climate change scenarios and/or models. The implementation of a comprehensive sensitivity analysis is, however, beyond the scope of this chapter and is recommended for possible future research. Refer also to Sect. 5.5 for a sensitivity analysis of the flexibility value of alternative coastal management strategies to different future conditions.

The method adopted in this chapter provides a way to use non-stationary probabilistic data for climate-proofing flood risk infrastructure. Perhaps the only major theoretical drawback of the proposed method is that it assumes probabilities can be given to future rainfalls under climate change (Fig. 4.2); many climate scientists do not believe this is yet possible. Nevertheless, the UKCP09 projections provide probabilistic data on future rainfall, which is conditional on the high, medium and low climate change scenarios. As no information is available on the likelihood associated with the climate change scenarios, equal probabilities have (subjectively) been assigned here to these scenarios.

Additionally, the usefulness of RIO analysis is limited by its assumption that the drift rate (i.e., the average magnitude of climate change per time unit) is constant. For some climate change scenarios, such as an approximately linear change over the century, this is an acceptable assumption—as in the case example presented in this chapter. For other climate change scenarios, such as rapid early changes or rapid later changes, it is, however, inappropriate to assume a constant drift rate. If the drift rate varies with time, then a different model (than GBM) is needed to describe the stochastic process for climate change.

# 5. Comparing Real In Options and Net Present Value

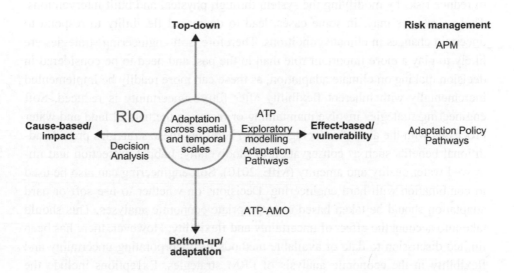

This chapter is based on:
Gersonius, B., Morselt, T., Nieuwenhuijzen, L. van, Ashley, R. & Zevenbergen,
C. 2011. How the failure to account for flexibility in the economic analysis of
flood risk and coastal management strategies can result in maladaptive decisions.
*Journal of Waterway, Port, Coastal, and Ocean Engineering* 1(96).

## 5.1. Introduction

Climate change has introduced large uncertainties into the assessment and management of flood risk. These uncertainties make it difficult to decide how to devise flood risk management (FRM) strategies and which measures (either single or portfolio) to use. In particular, it is widely recognised that there is a need to succeed (quasi-)stationarity-based procedures for developing FRM strategies (Kundzewicz et al., 2008). Otherwise, such strategies can be maladaptive, resulting in unnecessary costs of potentially irreversible measures (Barnett and O'Neill, 2010). This is particularly significant for hard engineering strategies, which aim to reduce risks by modifying the system through physical and built interventions. These strategies may, in some cases, lead to decreased flexibility to respond to uncertain changes in climate conditions. Therefore, soft engineering strategies are likely to play a more important role than in the past and need to be considered in decision making on climate adaptation, as these can more readily be implemented incrementally with inherent flexibility after future uncertainty is reduced. Soft engineering strategies involve maintaining or restoring the natural land and water processes with the aim of reducing risks. In addition, these strategies provide additional benefits such as conservation of biodiversity, habitat protection and improved water quality and amenity (MfE, 2010). Soft engineering can also be used in combination with hard engineering. Decisions on whether to use soft or hard adaptation should be taken based on appropriate economic analyses. This should take into account the effect of uncertainty and flexibility. However, there has been limited discussion to date of available methods for incorporating uncertainty and flexibility in the economic analysis of FRM strategies. Exceptions include the studies by Wang and de Neufville (2004), Gersonius et al. (2010), Woodward et al. (2011) and De Bruin and Ansink (2011).

The aim of this chapter is to analyse how the failure to incorporate uncertainty and flexibility in the economic analysis of FRM strategies can, in some cases, result in maladaptive decisions by using two alternative, but complementary economic analysis methods: Net Present Value (NPV) and Real In Options (RIO). RIO analysis offers a significant extension of the conventional NPV method, because it integrates expected changes in future levels of uncertainty into economic analysis. This method has been applied to the semi-hypothetical case study of a coastal defence system in order to demonstrate its applicability for decision making on climate adaptation. However, the results are not limited to coastal defence, and it would also have been possible to develop a flood defence, drainage or other water example. In the case study, two different FRM strategies are analysed: defence raising (i.e. dike heightening) and sand nourishment (i.e. the placement of sand in front of the dike). The former comprises the hard alternative and the latter the soft alternative.

## 5.2. Methods

### Economic analysis without uncertainty and flexibility

Conventional economic analysis of FRM strategies generally includes, for each alternative strategy, a calculation of its NPV. This term is used to describe the sum of the discounted benefits of an alternative less the sum of its discounted costs, all discounted to the same base date (e.g., HM Treasury, 2003). In this chapter, a negative NPV is referred to as a Net Present Cost (NPC) (*ibid*)—this is because it considers only costs, not benefits. NPV analysis allows the comparison of alternative FRM strategies with different patterns of benefits and costs over time, because it converts all benefits and costs into a single value at the base date. In calculating the NPV, the most likely values of uncertain variables are incorporated into the estimation of the benefits and costs. This should be analysed over the same time horizon for all alternatives. If a full benefit cost analysis has been undertaken, then the decision rule is to select the alternative that maximises NPV. In a cost effectiveness analysis, as applied in this chapter, the decision rule is to select the alternative that minimises NPC.

There are unfortunately two major limitations of the conventional NPV method. Firstly, the method is based on expectations of future investments (assuming e.g. an average or worst-case scenario). There may, however, be other (more extreme) scenarios where the life cycle cost will be different from expectations. Secondly, it uses a deterministic investment path for the static adaptive strategy. The working assumption is that the adaptive strategy continues unchanged until the end of the time horizon. This reasoning neglects the effects that management decisions may have under the extremely low or extremely high scenarios, because it assumes commitment by decision makers to a certain investment path. Consequently, the conventional NPV method does not adequately reflect the flexibility that exists in alternative adaptive strategies. It is of note that some more complete methods to dealing with uncertainty could partly address these limitations. For example, the use of NPV analysis in combination with Monte Carlo simulation could provide information on the life cycle cost of the alternative FRM strategies across the range of uncertainties. However, this cannot properly quantify the value of managing uncertainty and flexibility.

**Economic analysis with uncertainty and flexibility**

RO analysis is a recognised procedure to handle uncertainties in infrastructure investments by providing managerial flexibility (Myers, 1984). Instead of assuming a deterministic investment path as in the NPV method, RO analysis is able to deal with the possibility of many alternative investment paths through time. It explicitly considers combinations of possible investment decisions. In this regard, it is an extension of the NPV method. RO analysis determines the value of flexibility within a framework that builds on (but does not apply) the financial options theory of Black and Scholes (1973). RO analysis has more recently been applied to the design of technical/infrastructure systems, which has been termed RIO analysis (De Neufville, 2003). Unlike conventional RO analysis, RIO analysis embeds real options directly into the engineering system (re)design. The application of RIO analysis, therefore, requires extensive knowledge about the technical/infrastructure system. The crux of RIO analysis lies in the estimation of the value of flexibility built into technical/infrastructure systems. This is because the estimation of the cost of acquiring flexibility is relatively simple; that is, it is part of the set of conventional economic analyses. The assessment of the value of flexibility is the novel part that requires additional procedures. These are (Wang and de Neufville, 2004):

> Estimating the drift and volatility of the uncertain variable. The drift is the average rate at which the uncertain variable changes and the volatility is a measure of its randomness.

> Using the drift and volatility to develop a path-dependent tree representation of the different possible future paths followed by the uncertain variable.

> Quantifying the value of flexibilities built into the engineering system (re)design using the tree of the uncertain variable as well as the cost functions for adapting the system.

## 5.3. Case study description

The case study is typical of the Helderse, Pettemer, Hondsbossche and Westkappelsche coastal defences (the Netherlands), where a single sea dike is in place to protect an area of low lying land from flooding. Dutch coastal defences have a protection standard of 1/10000 years; i.e. they are designed for a tidal event with a probability of occurrence of 10-4. Overtopping of the defences is assumed to be the critical failure mechanism. The defences should be high (crest) and strong (inner slope) enough to resist a design overtopping volume of $q=1$ l/m/s at the hydraulic peak conditions.

The hydraulic load on the sea dike comprises of the overtopping discharge caused by the combination of the design water level and wave run-up. In the semi-hypothetical example, the design water level with a probability of occurrence of 10-4 equals NAP (Amsterdam Ordnance Datum) +5.0 m. The significant wave height accompanying the design water level is approximately 3.5 m with a steepness of 3.0 % and a period of 8.6 s. In addition to the hydraulic parameters, the overtopping discharge is primarily determined by the defence crest height. Other elements that influence the overtopping discharge are a gentle outer slope, a wide outer berm and/or a rough revetment. The structure has a total length of 10.0 km. The crest height is circa NAP +12.0 m with a width of 3.0 m. The outer side of the sea dike has a slope above the berm (located at NAP +5.0 m) of 1:3 and below the berm of 1:4. The inner side of the sea dike consists of a 1:3 slope and an inner berm, with maintenance road and ditch.

Predicted accelerating sea level rise (SLR) as a consequence of climate change will increase the loading on the sea dike, such that the system performance progressively deteriorates over time. This means that there is a (recurring) need to adapt the structure to comply with the protection standard. Two potential FRM strategies will be discussed in the following. These strategies will be briefly explained and some preliminary calculations will be made for the required adaptation and costs for the measures needed to withstand the hydraulic peak conditions in future time periods.

**The hard structural alternative**
The hard structural alternative comprises the continuation of the current FRM strategy, which aims to meet the protection standard by simply raising the sea dike (Fig. 5.1). In this case, the structure has to be strengthened in the landward direction by broadening the base. This requires a wider footprint of the dike at the landward side, which is some 6 m extra width per 1 m of dike heightened. Secondly, the infrastructure at the inner toe has to be relocated; i.e. the maintenance road and ditch. The inner slope of the sea dike comprises a layer of clay covered with grass. Where there is significant dike heightening, the existing clay layer first has to be removed in order to prevent the inclusion of sand between the clay layers. The required adaptation to the crest level was determined based on its relation with SLR, using Hydra K (Veugdenhil et al., 2000) and the PC-Overtopping tool (TAW, 2002). Hydra K is a probabilistic model to derive representative hydraulic conditions for coastal areas in the Netherlands. PC-Overtopping is an empirical model to make preliminary predictions for overtopping discharges for dike type structures. The capital cost of raising the sea dike to continue to maintain the protection standard in the face of climate change, was estimated based on unit cost prices from previous studies for the Dutch North Sea coast (Van Koningsveld, 2004). The outcome of this analysis is presented in Table 5.1. It can be concluded from Table 5.1 that the capital cost estimates change almost linearly with

the magnitude of SLR, with the following cost function (Fig. 5.2): evolution cost [M€] = 17.00 * magnitude of SLR [m] + 29.33. The marginal annual maintenance cost of defence raising is very low (*ibid*), and set at zero.

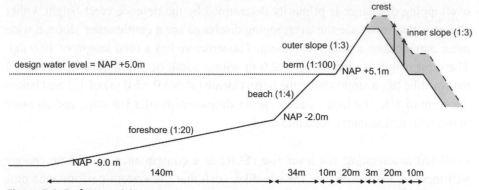

**Figure 5.1.** Defence raising

**Table 5.1.** Indicative capital cost estimates of defence raising

| SLR | Required crest level | Defence raising | Extra required footprint | Capital cost |
|---|---|---|---|---|
| [m] | [m +NAP] | [m] | [m] | [M€] |
| 0.50 | 12.66 | 0.66 | 3.98 | 38.0 |
| 1.00 | 13.35 | 1.35 | 8.11 | 46.0 |
| 1.50 | 14.04 | 2.04 | 12.24 | 55.0 |

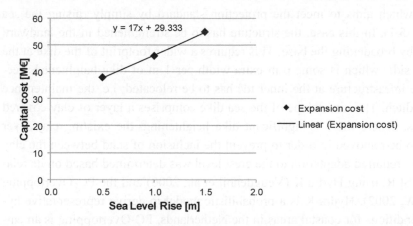

**Figure 5.2.** Capital cost of defence raising

## The soft structural alternative

The soft structural alternative comprises the placement of sand in front of the sea dike to maintain a higher foreshore level (Fig. 5.3). An elevated foreshore reduces the energy of waves through the action of the added resistance to run-up and by causing the waves to break before reaching the dike. This can reduce the overtopping volume, which has a beneficial effect on the required crest level. In this regard sand nourishment can help to avoid the need for dike heightening in the (near) future. The nourishment volume is calculated from the site area and height required. The part of the foreshore between NAP –9.0 m and the dike toe (located at NAP –2.0 m) has a 1:20 slope, and in the deeper parts, the slope is 1:10. Based on expert opinion, the foreshore length is taken to be about 0.5 times the wave length, which comes to about 50 m. The required foreshore height is determined as a function of SLR, based on the existing dike crest level (NAP +12.0 m). The outcome is presented in Table 5.2, along with the associated nourishment volume. The unit costs of nourishment are estimated to be 3 Euro per m3 for foreshore nourishment and 6 Euro per m3 for beach nourishment, after Morselt (2009). By applying these unit cost prices, the resulting initial capital cost estimates for sand nourishment are shown in Table 5.2. This gives the following linear cost function (Fig 5.4): initial capital cost [M€] = 8.04 * magnitude of SLR [m] + 13.38. The cost function for expanding the initial design of the foreshore is then: evolution cost [M€] = 8.04 * magnitude of SLR [m]. It can be seen from this that the capital costs of sand nourishment are lower than of implementing defence raising. However, replacing the sand as it is washed away requires annual maintenance. The costs for this are estimated to be approximately 10 % of the total nourishment volume (*ibid*).

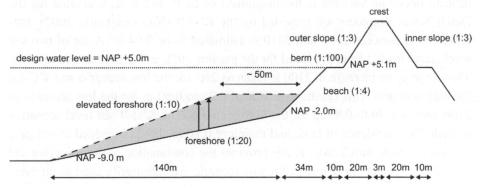

**Figure 5.3.** Sand nourishment

**Table 5.2.** Indicative capital cost estimates of sand nourishment

| SLR | Required foreshore height | Required foreshore nourishment volume | Required beach nourishment volume | Total capital cost |
|---|---|---|---|---|
| [m] | [m +NAP] | [m³/km] | [m³/km] | [Million €] |
| 0.50 | 0.50 | 245,000 | 168,000 | 17.4 |
| 1.00 | 1.50 | 245,000 | 235,000 | 21.4 |
| 1.50 | 2.50 | 245,000 | 301,500 | 25.4 |

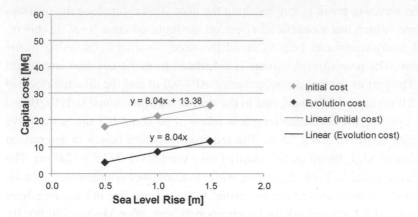

**Figure 5.4.** Capital cost of sand nourishment

## 5.4. Application

### Economic analysis without uncertainty and flexibility

The application of the NPV method requires the estimation of size and timing of investments within the system lifetime. The alternative FRM strategies should be defined in advance based on a specified scenario for the most significant uncertain variable in order to obtain the investments. In the case study, the most significant uncertain variable is the magnitude of SLR. Sea level scenarios for the Dutch North Sea coast are provided by the KNMI (Van den Hurk, 2007). Observed SLR between 1990 and 2010 is estimated to be 0.04 m. A set of two sea level scenarios has been produced for the periods 2050 and 2100, relative to 1990. The temperature increase in 2100 is taken as 2°C for the low scenario and 4°C for the high scenario. This results in a SLR of 0.35 to 0.60 m for the low scenario in 2100, and of 0.40 to 0.85 m for the high scenario. The KNMI sea level scenarios exclude the subsidence of land, and therefore 0.10 m should be added to estimate the relative SLR until 2100, i.e. this provides the combination of sea level rise and subsidence. The upper bound of the low scenario was arbitrarily used as an average scenario to define the required adaptation. This gives a relative SLR of 0.65 m between 2010-2100.

Any economic analysis should consider how the increase in flood risk/probability due to SLR is managed over time: i.e., either with a static robust strategy or with a dynamic adaptive strategy (Fig. 5.5). Under the static robust strategy, with a one-off adaptation step, the flood risk/probability decreases sharply at the outset of the project and then increases over time towards the level of acceptable risk (which is assumed constant). Under the dynamic adaptive strategy, with multiple adaptation steps, the level of risk/probability follows a saw-tooth pattern over the system lifetime (Fig. 1.1). The dynamic adaptive strategy will be appropriate in the majority of cases (for both hard and soft engineering strategies), because it will lead to economically optimal adaptation steps. This (i.e., the step size) will be dependent on the economics of fixed and variable costs of the specific alternative under consideration (refer to Fig. 5.2 and 5.4). In some cases, however, it will be necessary to adopt a static robust strategy—for example, when taking multiple adaptation steps over the time horizon is too complex to manage or not acceptable to society or for the environment. For purposes of illustration, the static robust strategy has been selected here for dike heightening, while the dynamic adaptive strategy has been selected for sand nourishment. This could potentially be justified by the fact that heightening the dike twice is unacceptable to the local community.

The valuation for dike heightening based on the static robust strategy is straightforward. Applying the cost function from Fig 5.2, the NPC for dike heightening is €40.38 million. The valuation of sand nourishment is somewhat more involved. It requires the planning of appropriate investment timings, and then discounting back these investments at the real discount rate of 5.5% (Financiën, 2009). The spreadsheet model for analysing the NPC of sand nourishment is shown in Table 5.2, assuming (arbitrary) adaptation steps of 15 years. The resulting cost for sand nourishment is €42.84 million. This implies that, as having the lowest NPC, dike heightening would likely be selected for implementation.

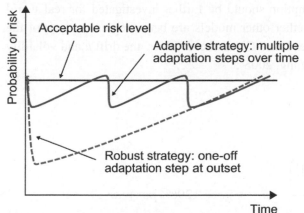

**Figure 5.5.** Strategies for adapting to climate change

Table 5.3. Spreadsheet model for analysing the NPC of sand nourishment

| Time period | Capital cost | AM cost | PC |
|---|---|---|---|
| | [M€] | [M€] | [M€] |
| t=0 | 14.13 | 1.41 | 29.10 |
| t=1 | 0.75 | 1.49 | 7.40 |
| t=2 | 0.82 | 1.57 | 3.50 |
| t=3 | 0.96 | 1.67 | 1.67 |
| t=4 | 0.96 | 1.76 | 0.79 |
| t=5 | 0.96 | 1.86 | 0.37 |
| NPC | | | 42.84 |

## Economic analysis with uncertainty and flexibility

As an initial step in the RIO analysis, the drift and volatility of the uncertain variable (i.e. the magnitude of SLR) were estimated from expert opinion. The expert (on climate change scenarios) gave an optimistic estimate for absolute SLR between 2010-2100 of 0.31 m, and a pessimistic estimate of 0.81 m, both with 90% confidence. Therefore the mean value over 90 years is 0.56 m and the standard deviation is 0.195 m, assuming the magnitude of SLR has a lognormal distribution. The lognormal distribution of SLR means the logarithm of the SLR has a normal distribution (i.e., it follows a generalized Wiener process, as explained below). This is a somewhat arbitrary assumption used for analytical convenience, and the validity of this assumption should be further investigated for real world case studies to determine whether other models are better representations of the stochastic movement of the sea level. Given these values, the drift $\mu$ and volatility $\sigma$ are calculated as follows (Wang, 2005):

$$\mu = \frac{\left(\dfrac{5.56-5}{5}\right)}{90} = 0.124\% \; per \; year$$

$$\sqrt{90}\sigma = \frac{\ln\left(\dfrac{5.56+(1.65\times0.195)}{5.56}\right)}{2} \quad or \quad \sigma = 0.296\% \; per \; year$$

The volatility above is calculated from the 95% confidence value, which is equal to the mean plus 1.65 times the standard deviation.

The evolution of the absolute SLR over time has been modelled by means of a path-dependent binomial tree representation (Wang and de Neufville, 2004). The binomial tree arises from a discrete random walk model of uncertainty (e.g., Wiener, 1923). This breaks down the time horizon into a number of time periods, or adaptation steps. The tree of the SLR uncertainty is then developed moving from the present to the end of the time horizon. According to the binomial tree, the sea level can only move upwards or downwards within each time period by a fixed factor. There is a specific probability of the up movement and down movement. Two methods are commonly used to develop the binomial tree: either the probabilities of the up or down movement are taken as equal and formulae are derived which give different up and down factors, or the uncertain variable is made to move up or down by the same factor, in which case formulae are derived which give different probabilities for those movements. With a sufficiently large number of time periods, these two methods converge on a single value. Here, the method has been used with equal probabilities, known as the Jarrow-Rudd binomial tree (Jarrow and Rudd, 1982). In this method, the up and down factors are calculated using the drift $\mu$, the volatility $\sigma$, and the time period $t$ as follows:

$$up\ factor = e^{\left(\mu - \frac{1}{2}\sigma^2\right)t + \sigma\sqrt{t}}, \quad down\ facto\ r = e^{\left(\mu - \frac{1}{2}\sigma^2\right)t - \sigma\sqrt{t}}$$

A binomial tree of absolute SLR with six time periods of 15 years has been developed. The reason for selecting a 15-year time period is that it typically takes one or more decades before a 'signal' of accelerating SLR can be detected in the observed sea level data. The binomial tree represents the different possible future paths of SLR uncertainty during the time horizon. A path-dependent binomial tree with six time periods results in 64 future paths to consider. The resulting probability density function of the absolute SLR at the end of the time horizon is shown in Fig. 5.6.

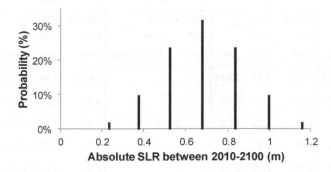

**Figure 5.6.** Probability density function of absolute SLR between 2010-2100

The way in which the effects of SLR uncertainty are dealt with over time will depend on the strategy used for adapting to climate change. The static robust strategy, associated with a one-off adaptation step, can only deal with the full range of uncertainty by preparing for the worst-case path of SLR. This strategy is selected for dike heightening, due to (supposed) social constraints on taking a dynamic adaptive strategy for this alternative. As such, the NPC of dike heightening is analysed for the worst-case path of SLR. This gives a higher NPC of €47.69 million, as opposed to the result of €40.38 million without any consideration of uncertainty. The reason for the higher NPC is that the defence is built higher initially than otherwise designed in the economic analysis without uncertainty, since adaptation in the future is not expected to happen. The dynamic adaptive strategy is selected for sand nourishment. This type of strategy allows the flexibility to manage future uncertainties by changing the engineering system (re)design as knowledge advances. This implies that the effects of the various ways to provide flexibilities need to be incorporated for this alternative by using RIO analysis. Table 5.4 shows the spreadsheet model to analyse the flexibilities within the design of the foreshore. The model incorporates the effects of the various ways to build in RIOs by changing the capital and maintenance costs to reflect the different possible design alternatives. The costs of the design alternatives are calculated with the help of RIO analysis. RIO analysis averages the NPC of the set of design alternatives over all possible future paths of SLR based on the probabilities derived from the stochastic process in order to obtain the Expected Net Present Cost (ENPC). The ENPC obtained for the sand nourishment is €44.84 million. These outputs show that sand nourishment is preferable to dike heightening when uncertainty and flexibility are incorporated into the analysis (as €44.84 million < €47.69 million).

**Table 5.4.** Spreadsheet model for analysing the ENPC of sand nourishment

| | Path | | | | | | | | | | | | |
|---|---|---|---|---|---|---|---|---|---|---|---|---|
| | p=0 | | | p=1 | | | p=... | | | p=63 | | |
| Time period | Capital cost | AM cost | PC | Capital cost | AM cost | PC | Capital cost | AM cost | PC | Capital cost | AM cost | PC |
| | [M€] | [M€] | [M€] | [M€] | [M€] | [M€] | [M€] | [M€] | [M€] | [M€] | [M€] | [M€] |
| t=0 | 14.73 | 1.47 | 30.32 | 14.73 | 1.47 | 30.32 | ... | ... | ... | 14.73 | 1.47 | 30.32 |
| t=1 | 1.38 | 1.61 | 8.30 | 1.38 | 1.61 | 8.30 | | | | 0.42 | 1.51 | 7.38 |
| t=2 | 1.42 | 1.75 | 4.03 | 1.42 | 1.75 | 4.03 | | | | 0.42 | 1.56 | 3.40 |
| t=3 | 1.46 | 1.90 | 1.95 | 1.46 | 1.90 | 1.95 | | | | 0.42 | 1.60 | 1.56 |
| t=4 | 1.50 | 2.05 | 0.94 | 1.50 | 2.05 | 0.94 | | | | 0.42 | 1.64 | 0.72 |
| t=5 | 1.55 | 2.20 | 0.45 | 1.55 | 2.20 | 0.45 | | | | 0.43 | 1.68 | 0.33 |

| NPC | 45.98 | 45.98 | | 43.71 |
|---|---|---|---|---|
| Prob | 0.016 | 0.016 | ... | 0.016 |
| **ENPC** | | | | **44.84** |

## 5.5. Sensitivity analysis

A sensitivity analysis has been carried out to analyse the effect of the discount rate and the volatility of the SLR rate on the choice of the FRM strategy. The reason for focusing on the volatility of the SLR rate (as opposed to the drift) is that any flexibility built into alternative FRM strategies is more valuable when there is higher volatility. This does not count for, or counts to a lesser extent for, higher drifts (i.e., higher average SLR rates). The results of the sensitivity analyses for both economic analysis methods (NPV and RIO) are shown in Fig. 5.7 and 5.8.

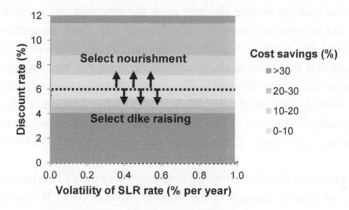

**Figure 5.7.** Sensitivity analysis of the investment decision, according to the NPV method

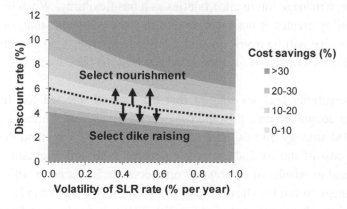

**Figure 5.8.** Sensitivity analysis of the investment decision, according to RIO analysis

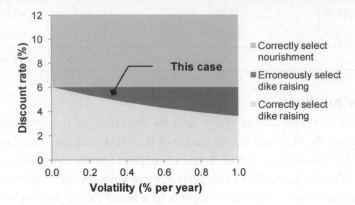

**Figure 5.9.** Possibility of erroneous decisions based on the NPV method

In the NPV method, the choice of the FRM strategy is sensitive to variation in the discount rate only. The best alternative changes once the discount rate exceeds 6%. If the discount rate is below 6% then dike heightening will likely be preferable, and if the discount rate is above 6 % then sand nourishment will likely be preferable. This means that the decision uncertainty concerning the best alternative is highest for a discount rate of about 6%. The degree of decision uncertainty reduces as the cost savings associated with the best alternative increase. In the NPV method, the volatility of the rate of SLR has no effect on the decision as to which strategy to use or on the degree of uncertainty associated with this decision.

In RIO analysis, the choice of the FRM strategy is sensitive not only to variations in the discount rate, but also to changes in the volatility rate of SLR. It can be concluded from these results that the relative cost of sand nourishment decreases as the amount of climate change uncertainty increases. This is because sand nourishment is better able to manage future uncertainties as it has flexibility. When the value that this flexibility creates is not incorporated into the economic analysis, the cost of this alternative will be overestimated. Only where there is no climate change uncertainty does RIO analysis give the same result as the NPV method.

The results of the sensitivity analysis suggest that the method selected for the economic analysis of adaptation can, in some cases, have a significant effect on the choice of the FRM strategy. In this extreme case study, the NPV method decreases the relative cost of the hard alternative compared with soft alternative. This, in turn, may lead to misplaced or at worst erroneous decisions as to which alternative FRM strategy to use (as shown in Fig. 5.9), and this is an example of maladaptation to climate change (Barnett and O'Neill, 2010). It can be seen from Fig. 5.9 that the possibility of erroneous decisions based on the conventional NPV method is highest in those cases where there is both high climate change uncertainty and high decision uncertainty concerning the best alternative.

## 5.6. Discussion and conclusions

Two methods to the economic analysis of adaptation were considered: NPV and RIO. The key difference between the two methods concerns the treatment of uncertainty and flexibility. While the conventional NPV method assumes a deterministic investment path, and does not incorporate the value of flexibility into the analysis, RIO analysis is able to deal with the possibility of many different investment paths through time, and explicitly accounts for the value of flexibility. In this sense, RIO analysis offers a significant extension on the conventional NPV method. RIO analysis has been used in this chapter for the context of a coastal defence system in order to support the choice between two alternative FRM strategies: dike heightening (the hard alternative) and sand nourishment (the soft alternative). For purposes of illustration, a static robust strategy has been adopted for dike heightening and a dynamic adaptive strategy has been adopted for sand nourishment. Using this example, it has been demonstrated how the failure to incorporate uncertainty and flexibility into the economic analysis can, in some cases, result in maladaptive decisions. Based on these results, it is concluded that the use of RIO analysis can help to avoid maladaptation for FRM. This is particularly relevant where there is both high climate change uncertainty and high decision uncertainty concerning the best alternative. This conclusion should, therefore, not be generalised to all cases, as NPV and RIO will, in many cases, give the same results regarding the best alternative (e.g., where there is no/low decision uncertainty).

It should, furthermore, be noted that the case study represented an extreme example, in which the flexibility value of dike heightening was equal to zero. This was because a static robust strategy has been adopted for dike heightening, which required the sea dike to be initially designed to accommodate the worst case path of SLR. In the majority of cases, however, the dynamic adaptive strategy would apply to any FRM strategy, both hard and soft. This means that the value of flexibility in each alternative FRM strategy should be taken into account in the economic analysis. In this regard, it is also important to note that (contrary to common presumption) the flexibility value of sand nourishment was quite limited in this case: the ENPC with learning was only 1.14 million Euro (or 2.5 %) lower than the NPC under the worst case path of SLR. This limited flexibility value could be explained by the large required foreshore volume initially, the high maintenance cost and the high discount factor.

Although RIO analysis can help to address the limitations of the conventional NPV method by extending it, it also has a number of limitations. A theoretical limitation is that it assumes probabilities can be given to future SLR under climate change; many climate scientists do not believe this is yet possible. A practical limitation is that it can be complicated to establish and then solve the binomial tree. In particular, the representation of the stochastic process for the SLR over time is a major challenge. In this case study, the SLR has been modelled as a random walk process characterised by a constant mean and standard deviation. In this model, all future paths for SLR were equally probable. This is not necessarily the best model for SLR: a linear model with an uncertain slope, combined with a random walk process, could potentially improve the analysis (Vrijling, personal communication). This is recommended for further research.

# 6. Adaptation Tipping Point - Adaptation Mainstreaming Opportunity

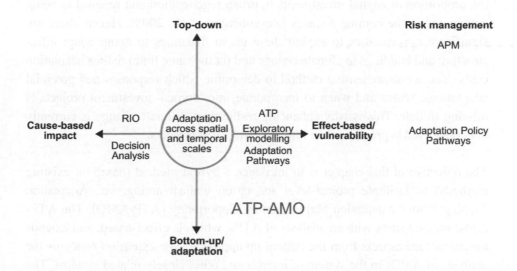

This chapter is based on:

Gersonius, B., Nasruddin, F., Ashley, R., Jeuken, A., Pathirana, A. & Zevenbergen, C. 2012. Developing the evidence base for mainstreaming adaptation of stormwater systems to climate change. *Water Research*, accepted.

## 6.1. Introduction

Adaptation to climate change is usually assumed to require additional financial capacity to better deal with more severe climate conditions. The implementation of adaptive strategies at the local level is, however, constrained by a lack of financial resources (in the short-term). As a consequence, enhancing resilience to climate change (or: climate proofing (Kabat et al., 2005) should as much as possible be based on the incorporation of adaptation responses with 'normal' investment projects, such as for the maintenance/modification/renewal of infrastructure, buildings and public spaces. In the Western world in particular, a steep increase in the proportion of capital investments in urban regeneration and renewal is being anticipated in the coming decades (Zevenbergen et al., 2008). Hence, there are significant opportunities to exploit these urban dynamics to better adapt infrastructure and buildings to climate change and (at the same time) reduce adaptation costs. Yet, a comprehensive method to determine which responses and potential adaptations, where and when to incorporate into 'normal' investment projects is missing to date. This is why enhancing resilience to climate change is currently not considered in practice when implementing such projects.

The objective of this chapter is to introduce a hybrid method (based on existing methods) to facilitate project-level adaptation mainstreaming: i.e., Adaptation Tipping Point - Adaptation Mainstreaming Opportunity (ATP-AMO). The ATP-AMO method starts with an analysis of ATPs, which is effect-based, and extends this to include aspects from the bottom-up approach. The extension concerns the analysis of AMOs in the system of interest and other closely related systems. The results from both analyses are then used in combination to take advantage of the right (i.e., cost-efficient) AMOs. Use of this method will enhance the understanding of the adaptive potential of the system. Adaptive potential refers to the ability of a system to adapt its structure and processes based on anticipated (re)developments within the assessment period (Lim et al., 2005), so that it will become more resilient to future change.

The remainder of this chapter elaborates the procedural steps of the ATP-AMO method and this is illustrated by a case study applied for flood risk in an urban stormwater system. Finally, the results of the case study application are discussed and conclusions drawn related to a wider range of objectives (e.g., pollution control) and the socio-economic changes accompanying climate change.

## 6.2. Method

The ATP method aims to assess the ability of a system to deal with climate (and other) change(s) by being robust/resilient enough to continue to function as required in the face of change. It, therefore, considers external drivers to see how large these can become before the system fails to function as required. The ATP method, as adapted from Kwadijk et al. (2010), involves the six steps below.

1. Start by specifying the functions and climate change effects of interest for the assessment. The objectives for these functions (which are often translated into acceptable standards) are also defined. In addition, the current strategy to achieve the objectives should be identified.

2. In the next step, the particular threshold values for the acceptable standards are quantified. These threshold values can be defined either according to regulations (e.g., by national law) or decided by the stakeholders involved and can change over time.

3. The ATPs are identified by increasing the design loading (e.g., the rainfall intensity) on the system, as a function of time, to assess the specific boundary conditions (i.e., the magnitude of climate change) under which acceptable standards may be compromised. This is much the same as a sensitivity analysis of the performance of the system to possible future design loadings. The results of the assessment are then represented in a bar chart, indicating the occurrence of ATPs (refer e.g. to Fig. 6.8).

4. Climate change scenarios are used to transform the specific boundary conditions (i.e., the magnitude of climate change) under which an ATP will occur into an estimate of when it is likely to occur. This can be done by overlaying these scenarios (in the above bar chart) with the possible future design loadings (refer e.g. to Fig 6.8). The output from this step will provide an estimate of the earliest and latest times that the performance of the stormwater system is likely to no longer be acceptable.

5. If it is desired that an ATP should not be reached, a change in/of the adaptive strategy will be needed to maintain or enhance climate change resilience. As there are long lead-in times to effecting structural measures this (i.e., the potential options for adapting the system) needs to be defined as early as possible and well before the critical ATP occurs.

6. Analysing the potential options for adapting the system and the ATPs (by repeating steps 3 and 4) will result in the definition of a number of adaptive strategies, some structural and some non-structural. Engagement with all stakeholders is required in this step to select an adaptive strategy that is realistic and acceptable. Implementing this strategy will alter the nature and timing of the critical ATPs. Alternatively, where the implementation of the adap-

tive strategy is too costly or not acceptable to society or for the environment, the acceptable risk level may be allowed to decline.

In the above method, the time window of an ATP will define when a change in the adaptive strategy will be needed. This assumes that climate change is the main driver of adaptation. In urban areas however, the maintenance/modification/renewal of infrastructure, buildings and public spaces could give an opportunity to reconsider the existing system from a different standpoint. Many adaptation responses can be effected synergistically with the cycles of maintenance, modification and renewal (e.g., Zevenbergen et al., 2007; Veerbeek et al., 2010) and at next to no additional cost (e.g., Van de Ven et al., 2011). From this perspective, these urban dynamics should be recognised and used as perhaps the most important driver of and opportunity for adaptation. Therefore, the ATP method has been extended here by including aspects of the bottom-up approach (after step 4 and 5 of the ATP method). The additional steps are to:

4A. Understand the cycles of maintenance, modification and renewal as well as the opportunities that these afford for the mainstreaming of climate adaptation of the system of interest. A simple but practical way to do this is to make use of predictions of expected physical lifetimes, as determined from expert knowledge and/or literature (Langston, 2008); although more sophisticated methods, like deterioration prediction modelling for infrastructure and buildings, could also be used. Note that if the cycles of maintenance, modification and renewal do not present an opportunity for project-level adaptation mainstreaming, then the associated investment projects should not be included in the remainder of the analysis.

4B. Determine the time windows when AMOs will occur. This can be done by estimating when the existing structures will reach their end-of-life, by considering their expected physical lifetimes (as obtained from step 4A) in relation to their construction periods (e.g., Veerbeek et al., 2010). Alternatively where, for example, neighbourhood regeneration is concerned, the time windows of AMOs can be directly obtained from the existing plans for the 'normal' investment projects. For the analysis of AMOs it is crucial that all the major stakeholders in the system share their plans for 'normal' investment projects as well as the timing of these.

4C. Modify the 'normal' investment projects to incorporate potential options for adapting the system. These options should then be include in the definition of the adaptive strategies in step 5 of the ATP method. Taking account of the possibilities for project-level adaptation mainstreaming in step 6 of the ATP method will lead to a better understanding and quantification of the adaptive potential of the system.

5A. Analyse the time windows of the AMOs and critical ATPs in conjunction. If the AMOs are likely to arise earlier than the critical ATPs, then it could be economically worthwhile to move the potential adaptation options (as identified in step 5) forward in time, so as to incorporate them into the 'normal' investment projects. As argued by Van de Ven et al. (2011), the costs of effecting adaptation responses synergistically with 'normal' investment projects will, in the majority of cases, be of the order of 50 to 80 % lower than the costs of implementing these responses as stand-alone adjustments. Whether (or not) project-level adaptation mainstreaming is likely to be cost-efficient will, however, also depend on the length of the differential time period between the occurrence of the AMO and the critical ATP. With a longer differential time period, the potential cost savings from adaptation mainstreaming will be off-set by the cost savings from postponing the implementation of adaptation responses until later, i.e., until the occurrence of the critical ATP (minus the lead-time). This is because later investments will be discounted more heavily than earlier investments. Hence, the longer the differential time period, the less attractive adaptation mainstreaming will be. Fig. 6.1 shows the percentage of cost savings (say X % more) that is required for cost-efficient adaptation mainstreaming. It is, finally, of note that when the AMOs are likely to arise later than the critical ATPs, then stand-alone adjustment will be required anyway instead of, or in addition to, adaptation mainstreaming. This could also imply implementing temporary responses in an attempt to delay the occurrence of the critical ATP until the AMO.

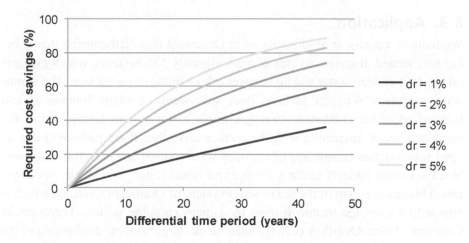

**Figure 6.1.** Required cost savings (≥ X %) for cost-efficient adaptation mainstreaming as a function of the differential time period between the occurrence of the AMO and the critical ATP and the discount rate

A flow chart of the ATP-AMO method is presented in Fig. 6.2.

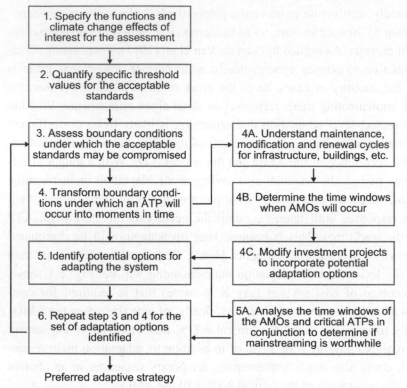

**Figure 6.2.** Flow chart of the ATP-AMO method

## 6.3. Application

Wielwijk is a post-war neighbourhood in Dordrecht (the Netherlands) that is being regenerated. It covers an area of approximately 135 hectares, with 4 hectares (about 3%) of open water and 40 hectares (about 30%) of paved area. The stormwater system in Wielwijk includes three subsystems: a minor drainage system (comprised of the combined sewers), a minor/major system (comprised of the combined sewer interacting with the surface and other flood pathways) and an open water system (comprised of the open watercourses and culverts). The combined sewerage collects sanitary sewage and urban runoff in a single piped system. This sewer is part of the larger sewer system of Dordrecht Centrum, which is pumped to a sewage treatment plant by a main pumping station. There are 34 Combined Sewer Overflow (CSO) outlets in the larger system discharging to the open water system, of which 6 CSOs are located in Wielwijk. Most of the open watercourses in Wielwijk lie at the edges of the neighbourhood (Fig. 6.3). These watercourses are fragmented, with a number of culverts.

**Figure 6.3.** Layout of the open water system; light grey lines represent the current situation and dark grey lines represent the future situation after the regeneration project has been implemented

The regeneration project in Wielwijk is occurring for various reasons, mainly social: an isolated and ill-used park, traffic nuisance, little diversity in neighbourhoods and water quality problems as a result of inadequate open water circulation. The municipality of Dordrecht and the housing corporation, in collaboration with residents, have proposed a plan for a neighbourhood regeneration project to address these problems. This plan is referred to as the Urban Vision 1.0, and is described below.

**Assessment of the system robustness**

In this case example, the function of interest is flood risk management while the climate change effect of interest is the change in rainfall intensity. In a fuller study the other hydrological effects of climate change such as mean seasonal rainfall, groundwater level and evapotranspiration rates might also be considered (Willems et al., 2011). These were deemed of lesser importance in this case study. In addition, a range of other objectives, such as avoiding unacceptable CSO pollution by controlling changes in urban runoff diffuse pollution caused by either climate change or implementing the responses for flood risk management, could also be included, adding additional ATPs and uncertainties for the analysis.

The objectives for flood risk management have been taken as the starting point for the analysis of ATPs (step 1). The existing stormwater system is required to be adequate to fulfil these objectives, performing to acceptable standards. The performance requirements are shown in Table 6.1, which sets out the acceptable standards, i.e., required capability in terms of flood frequency for the different subsystems, and their interaction (step 2). It can be seen from Table 6.1 that there are major differences in the acceptable standards for the different subsystems. The reasons for these differences are in the way the objectives have been formulated (refer to Fig. 6.4). The objective for the minor drainage system is to reduce water nuisance (i.e., water on the streets due to manhole flooding), while the objective for the minor/major drainage system is to minimise property damage from flooding. The minor/major system can be considered as a 'system within a system' for it includes two distinct but conjunctive drainage networks. The minor drainage system is designed for a 1 in 2 year event (which is specified in the Municipal Sewer Plan), but the whole system (minor/major) performance can be expected to be greater than this because exceedance flows are contained within surface pathways (e.g. road surfaces). RIONED, which is the centre of expertise in sewer management and urban drainage in the Netherlands, have proposed a standard for the minor/major system of 1 in 50 years (RIONED, 2006). The objective for the open water system is also to minimise property damage from flooding. In contrast with the minor drainage system, the open water system has generally no headroom in terms of the residual performance of the whole system over and above the capacity of the engineered system. Acceptable standards for open water systems have been established by national law as 1 in 100 years for urban areas (NBW, 2003). Within the context of the Water Plan Dordrecht, the City of Dordrecht and the Waterboard Hollandse Delta have set an additional objective to prevent manhole flooding as a result of interaction (of the minor/major system) with the open water system. This objective states that water level changes in the open water system exceeding 25 cm are allowable no more than once in 2 years, because of the relative elevation of the CSO discharge outlets. If the open water level rises above the minimum required elevation of the CSO discharge outlets, water levels in the minor/major system may be affected and hence its conveyance capacity may be reduced.

A 1-D/2-D Sobek Rural model was used for the investigation of the performance of the different subsystems and their interaction under the current and possible future design loadings, based on a fully integrated model of the subsystems. Sobek Rural is a numerical hydraulic modelling system to simulate hydrodynamics of one-dimensional channel flow and two-dimensional overland flow (Dhondia and Stelling, 2002). The 1-D Sobek model has been calibrated using measured water levels in the minor drainage system and open water system (Luijtelaar et al., 2006). It has been used to assess the specific boundary conditions (i.e., the magnitude of climate change) under which the system performance becomes unacceptable—that is, when an ATP will be reached (step 3). The model application is discussed in the following subsections.

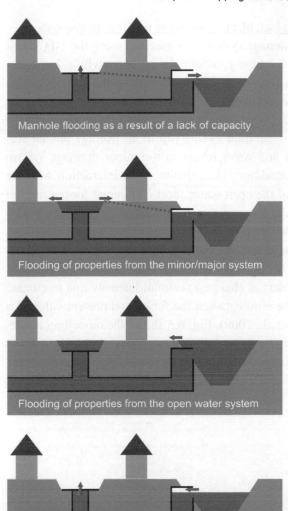

**Figure 6.4.** Illustration of the objectives for flood risk management

**Table 6.1.** Objectives and acceptable standards of flood risk management

| Objective | Annual frequency | Source |
| --- | --- | --- |
| Prevent manhole flooding as a result of a lack of capacity in the minor drainage system | 2 years | Municipal Sewer Plan |
| Prevent flooding of properties from the minor/major system | 50 years | RIONED (2006) |
| Prevent flooding of properties from the open water system | 100 years | NBW (2003) |
| Prevent manhole flooding as a result of inter-action with the open water system | 2 years | Water Plan Dordrecht |

Manhole flooding as a result of a lack of capacity in the minor drainage system

The performance of the minor drainage system was assessed using the 1-D Sobek model. The model was run with a 2-hour synthetic design event with a 2 year average annual frequency (20 mm rainfall in 120 minutes), i.e. the current design loading. This critical duration event has been obtained from the guideline for Dutch sewer systems (RIONED, 2004), which provides information on rainfall intensities for a number of short-duration events (up to an average annual frequency of 10 years). The flows and water levels in the minor drainage system were simulated for free outfall conditions (i.e., assuming no interaction with the open water system), as the rise of the open water level is minimal for the rainfall event used. The modelling results showed that no manhole flooding (at a 0.99 level of significance) occurs under the current design loading. Further simulations were undertaken for possible future design loadings by applying a series of climate change factors (cc factors) for rainfall intensities. A cc factor is a simple but useful method to describe the potential change in rainfall intensity due to climate change. It has been defined as the ratio between the future and present value of a hydro-climatic variable (Larsen et al., 2009). Fig. 6.5 shows the modelling results in terms of the resulting freeboard between the manhole water level and the street level, for current and possible future design loadings. According to these results, the minor drainage system performance becomes unacceptable for a cc factor of 1.25.

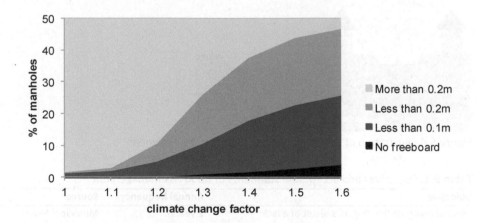

**Figure 6.5.** Simulated freeboard in manholes for the critical duration event

Flooding of properties from the minor/major system

The 1-D/2-D Sobek model was used to assess the whole system (minor/major) performance in terms of the passage of flows across urban surfaces. This model was run with a 1 in 50 year synthetic event with a critical duration of 2 hours (41 mm rainfall in 120 minutes), i.e. the current design loading. This event has been derived from the intensity-duration-frequency (IDF) curves of Buishand en Wijngaard (2007) (see Eq. 6.1 to 6.3) with the alternating block method (Chow et al. 2005).

$$x(T,D) = \lambda + \beta\left(\frac{1}{-0.09}\left(1 - T^{0.09}\right)\right) \tag{6.1}$$

$$\lambda = 1.618 + 1.862 \ln D + 0.274(\ln D)^2 \tag{6.2}$$

$$\beta = \left(1.168 + 1.862 \ln D + 0.274(\ln D)^2\right) * \left(0.242 + 0.124 \ln D - 0.022 \ln D^2\right) \tag{6.3}$$

Where $x(T,D)$ is the rainfall volume (in millimetres) of the storm event with an average annual frequency of $T$ years and a duration of $D$ minutes. The reason for using IDF curves is that only hourly rainfall time series are available for the Netherlands; while the assessment of the minor/major system performance requires rainfall time series in preferably five minute intervals, given the fast response time of the system. The 2-D overland flow module simulates the surface conveyance by the major system as well as the surface flooding. This module uses a high resolution (1x1 meters) Digital Elevation Model of the surface, including artificial features such as roads and buildings (Fig 6.6).

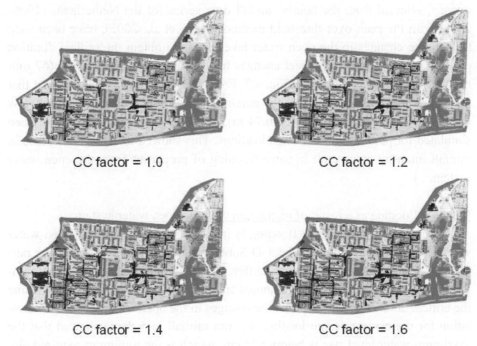

CC factor = 1.0    CC factor = 1.2

CC factor = 1.4    CC factor = 1.6

**Figure 6.6.** 1x1m DEM, showing the flood extent (in black) for the critical duration event

The 2-D overland flow module is fully integrated with the 1-D channel flow module to account for the capacity of the minor drainage system and the interactions between the minor and major drainage system. This approach allows the estimation of the extent and surface depth of flooding from the minor/major system (Fig. 6.6). It can be observed from Fig. 6.6 that no flooding of properties occurs for the current design loading. Simulations for possible future design loadings showed that some properties may be flooded when the present rainfall intensity is increased by 40%.

Flooding of properties from the open water system

The open water level, relative to the street level, governs the amount of hydraulic head available for drainage from the urban area, as it determines to what extent the open water levels may be lowered below the lowest street level (Schultz and Wandee, 2003). When the open water level rises above the lowest street level, flooding of properties may occur (as the urban area cannot then be drained). Wielwijk is an independent water level area (WK09) within the Weeskinderendijk polder. Pumps are used to control the water level in the polder, which has a preferred level of NAP (Amsterdam Ordnance Datum) -1.4 m in Wielwijk. The maximum pumping capacity of the Weeskinderendijk polder is 70 mm/day. The performance of the open water system was assessed with the 1D SOBEK model. Different duration historical rainfall events with a 50 year average annual frequency, selected from the hourly rainfall time series for the Netherlands (1906-2008) with the peak-over-threshold method (Chow et al., 2005), have been used to simulate changes in the open water level, so as to obtain the critical duration event. The simulated water level changes for the current design loading (67 mm rainfall in 5 days) are shown in Fig. 6.7. From this figure it can be observed that no properties will be flooded, as the maximum rise in open water level is below the 1% lowest street level (NAP -0.74 m). Water level changes have also been simulated for possible future design loadings. This showed that a 90% increase in rainfall intensity may result in some flooding of properties from the open water system.

Manhole flooding as a result of interaction with the open water system

A common cause of manhole flooding is insufficient capacity in the open water system (as explained above). The 1-D Sobek model was used to assess the performance of the open water system. Different rainfall durations for historical rainfall events with a 2 year average annual frequency were simulated to determine the critical duration event based on the changes in the open water level. The simulation for the current design loading (67 mm rainfall in 5 days) showed that the maximum water level rise is below +25 cm, which is the minimum required elevation of the CSO discharge outlets. Simulations were also performed for possible future design loadings. These results are presented in Fig. 6.8 in terms of water level changes. It can be observed from Fig. 6.8 that the open water system performance becomes unacceptable for a cc factor of about 1.05.

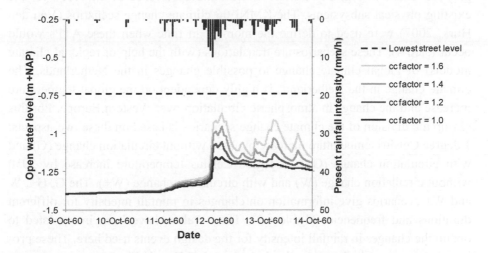

**Figure 6.7.** Simulated water level changes for the historical rainfall event from 9-14 October 1960

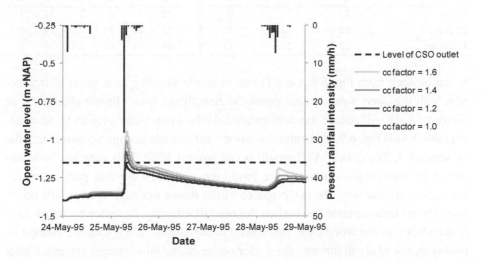

**Figure 6.8.** Simulated water level changes for the historical rainfall event from 24-29 May 1995

The occurrence of ATPs

The modelling results are represented graphically in Fig. 6.9.A-6.9.D in terms of the ATPs for the different subsystems and their interaction. The ends of the light grey bar charts in Fig. 6.9.A-6.9.D indicate the occurrence of the ATPs for the existing physical subsystems. The KMNI'06 climate change scenarios (Van den Hurk, 2007) were used to define the moment in time when these ATPs would occur (step 4). These scenarios are translations (with the help of regional climate models) of global climate change to possible changes in the Netherlands. The climate change in the Netherlands is highly dependent on the global temperature increase and the change in atmospheric circulation over Western Europe. For this reason the division of the climate change scenarios is based on these two aspects: 1 degree Celsius temperature increase by 2050 without circulation change (G) and with circulation change (G+), 2 degree Celsius temperature increase by 2050 without circulation change (W) and with circulation change (W+). The G, G+, W and W+ scenarios give information on changes in rainfall intensity for different durations and frequencies (Table 6.2). These values have been interpolated to obtain the changes in rainfall intensity for the design events used here. These projected changes are indicated by the dashed vertical lines in Fig 6.9.

**Table 6.2.** Projected changes in rainfall intensity in % from 1990 till 2050

| Frequency | 1 hour event | | | | 1 day event | | | | 10 days event | | | |
| --- | --- | --- | --- | --- | --- | --- | --- | --- | --- | --- | --- | --- |
| | G | G+ | W | W+ | G | G+ | W | W+ | G | G+ | W | W+ |
| 1 year | 7 | - | 21 | - | 9 | 6 | 18 | 9 | 6 | 1 | 11 | 3 |
| 10 years | 11 | - | 22 | - | 11 | 6 | 22 | 11 | 7 | 2 | 14 | 4 |
| 100 years | 12 | - | 23 | - | 11 | 6 | 24 | 11 | 8 | 2 | 15 | 5 |

It can be seen from Fig. 6.9.A-6.9.D that manhole flooding as a result of interaction with the open water system poses the first threat from climate change. A cc factor of 1.05 will reduce the performance of the open water system to an unacceptable level (Fig. 6.9.D). Under the worst case climate change scenario (i.e., the W scenario), this critical ATP would occur around 2020. This year has been obtained by interpolation between the base year and 2050 for this particular scenario. In a similar way, the performance of the minor drainage system will be reduced to an unacceptable level when the rainfall intensity increases by about 25% (Fig. 6.9.A). In the worst case, this would occur around 2055. Climate change is, however, not likely to threaten the performance of the minor/major system. Under the worst case climate change scenario, the whole system (minor/major) performance remains acceptable at least until about 2100 (Fig. 6.9.B). This means that this subsystem has a high degree of robustness.

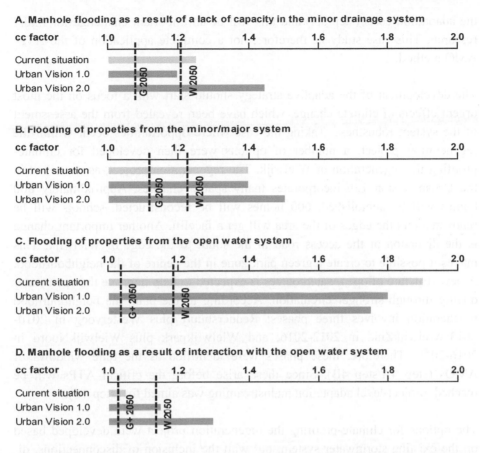

**A. Manhole flooding as a result of a lack of capacity in the minor drainage system**

**B. Flooding of propeties from the minor/major system**

**C. Flooding of properties from the open water system**

**D. Manhole flooding as a result of interaction with the open water system**

**Figure 6.9.** The occurrence of the ATPs (i.e., the ends of the bar charts) for the different subsystems and their interaction. To clarify: (i) the light grey bar charts represent the current situation; (ii) the dark grey bar charts represent the original and the modified plan for the neighbourhood regeneration; (iii) the dashed vertical lines show the KMNI'06 climate change scenarios for 2050

## Assessment of the adaptive potential

Once the robustness of the existing stormwater system has been defined, the (potential) next step is to understand and quantify its' adaptive potential. This requires analysing AMOs and, ultimately, taking advantage of the right (i.e., cost-efficient) AMOs. This so-called project-level adaptation mainstreaming can be seen as a no/low-regret strategy (on the condition that it does not entail major trade-offs or excessive costs) (Persson and Klein, 2009). A no/low-regret strategy is one where nil to moderate investment levels maintain or enhance resilience to climate change. Note that most adaptation options will be low-regret rather than no-regret, since, in practice, their incorporation into 'normal' investment projects will imply some additional costs. Here, there has not been any attempt to quantify the adaptation costs (step 5A) and this will be an important challenge for future research. This case study is, therefore, not a complete application of the ATP-AMO method.

The development of the adaptive strategy should start with a focus on the most urgent effects of climate change, which have been revealed from the assessment of the system robustness. Taking the case where opportunities occur within the regeneration project, a number of options were then developed for climate-proofing the regeneration of Wielwijk. The regeneration project, as described in the Urban Vision 1.0, incorporates many spatial changes. Approximately 800 homes will be demolished, 600 homes will be reconstructed, schools will be renovated and the edges of the area will get a facelift. Another important change is the diversion of the access road in the south to the edge of Wielwijk. This makes it possible to create a green park zone in the centre of the neighbourhood. A new structure of open watercourses is expected to help improve the open water quality through stronger circulation. According to the plan, the neighbourhood regeneration involves three phases: Reddersbuurt plus Westervoeg in 2010-2013;WielwijkZuid in 2012-2016; and Wielwijkpark plus WielwijkNoord in 2016-2019. Therefore, these phases were identified as the time windows of AMOs (step 4A/step 4B). Since these arise before the critical ATPs will be reached, project-level adaptation mainstreaming was aimed for (step 4C).

The options for climate-proofing the regeneration project were developed based on the existing stormwater system, but with the inclusion of disconnections, diversions and extra green and blue areas in the public space (e.g., Ashley et al., 2011; USEPA, 2010). These options were identified in a series of collaborative design workshops attended by urban designers, architects, sewer managers, water managers (including the water board), regeneration planners, scientists, and inhabitants. This collaborative design work was executed within and supported by a learning and action alliance (LAA) approach (Ashley et al., 2011). A discussion of how the LAA was organised to support the decision making on modifying the regeneration project is beyond the scope of this chapter; refer to Van Herk et al. (2011) for details. The (re)design challenge for the LAA was to identify adaptation options that have potential to maintain or enhance climate change resilience, while also being beneficial for urban heat island problems as well as providing amenity values and improvements to water quality, although no specific targets were set for these. The options selected can be grouped into those that concern either the minor drainage system or the open water system as summarised in the following. Through collaboration with the urban designers and architects, it has been possible to incorporate these options into the neighbourhood regeneration project (step 5). This is referred to as the Urban Vision 2.0.

The approach taken for the minor drainage system was to plan to reduce the volume of stormwater entering the combined sewer and to accommodate it elsewhere. This includes disconnecting roofed areas of public buildings and paved areas (such as streets and parking lots). The flows from most of these areas were diverted to the open water system via a new sewer for stormwater (as space was not available in the street profile for surface conveyance) despite the known risk that this could lead to future urban diffuse pollution problems (e.g., Mitchell, 2001). Where diversion to the open water system was impractical due to long distances, the stormwater was diverted to adjacent public green spaces. This required a lowering of the green spaces to allow for temporary surface storage, including underground storage. Such temporary surface storage provides spatial amenities for most of the time, and fills up in a controlled manner during intense rainfall events. This technique has been chosen for the park zone in the centre of the neighbourhood (Fig. 6.9). Using this approach, 28% of the total roofed area and 57% of the total paved area were considered redirected to the open water system and 7% of the total roofed area and 9% of the total paved area were considered to be completely disconnected. The latter volume of stormwater is stored locally (either on the surface or underground) before being released gradually into the sewer network.

The approach taken for the open water system was to raise the CSO discharge outlets and to provide additional open water (Fig. 6.10). The CSO discharge outlets were raised by 0.10 m to reduce manhole flooding as a result of interaction and backing up from the open water system. The extra open water amounted to about 1.9 ha, which equates to some 2% of the total area of Wielwijk. This provision of extra open water enhances the capacity of the open water system; is beneficial in taking the runoff from the disconnected roofed and paved areas from the minor drainage system; improves the open water quality through stronger circulation; and provides amenity value.

Finally, the adaptive potential of the stormwater system was quantified by analysing the ATPs for the modified plan for the neighbourhood regeneration (step 6). The resulting ATPs are shown by the ends of the dark grey bar charts in Fig. 6.9.A-6.9.D. It can be observed from Fig. 6.9.D that after the adaptive potential has been realised, the performance of the open water system will remain acceptable up to a 30% increase in rainfall intensity. According to the worst case climate change scenario (i.e., the W scenario), this ATP will not be exceeded until about 2100. Furthermore, the minor drainage system performance will become acceptable up to a 45% increase in rainfall intensity (Fig. 6.9.A). This ATP will occur around 2100 at the earliest. As a result of the increase in conveyance capacity of the minor drainage system, the whole system (minor/major) performance remains acceptable over an even longer time period (Fig. 6.9.B).

**Figure 6.10.** Illustration of the options selected for the minor/major system (bottom left) and open water system (bottom right). To clarify: (i) the bottom left figure shows the modified design for the park zone, which comprises a lowering of the green space to allow for temporary surface storage; (ii) the bottom right figure shows the design for the new open watercourse; (iii) the top photos represent the current situation at the two locations

## 6.4. Discussion and conclusions

Effect-based approaches for climate impact and adaptation assessment do not rely on precise forecasts of climate change. A relatively simple method for applying this approach is to use the concept of ATPs. This aims to assess the ability of a system to deal with climate (and other) change(s) by being robust/resilient enough to continue to function as required in the face of change. The ATP method has been successfully applied to the flood protection system of the Thames Estuary (Lowe et al., 2009) and the Netherlands water management system (Kwadijk et al., 2010). However, a common criticism to this method is that it leads to a reactive attitude on the part of the decision makers: "we will only act when it is necessary" (Jeuken and Te Linde, 2011). The latter is justified by the economical value of postponing infrastructure investment. Yet, there are significant opportunities arising from the maintenance/modification/renewal of infrastructure, buildings and public spaces to undertake adaptation, both now and in the future (Van de Ven et al., 2011; Veerbeek et al., 2010). Such opportunities occur particularly at the urban scale. The ATP-AMO method has been developed here in order to use the urban dynamics as a driver of adaptation, which is the novel contribution of this chapter.

It is of note that the adaptive potential will be different from the actual adaptive capacity, which is constrained by socio-economic, political and physical factors (Lim et al., 2005; Burton and May, 2004). Therefore, an important research initiative will be to study the adaptive capacity of urban stormwater systems (including the actors and rules). This will involve the specification and alignment with socio-economic scenarios, as these provide the context in which adaptation will take place.

The main finding of the case study is that the application of the ATP-AMO method helps to increase the no/low-regret character of adaptation (as defined in Sect. 6.3). This can be explained as follows:
> The assessment of the system robustness will improve the knowledge as to which responses and potential adaptations may be no/low-regret, as it focuses the attention on the most urgent effects of climate change;
> Project-level adaptation mainstreaming is expected to lead to potential cost savings, since adaptation responses can be incorporated with "normal" investment projects instead of being applied separately. For example, Van de Ven et al. (2011) claim that the majority of responses linked to 'normal' investment projects can be achieved at almost no additional cost;
> The collaborative design work will lead to the development of area-specific responses, which could not have been developed on a higher scale level. Often, such responses can keep climate adaptation costs down and/or achieve a range of wider benefits such as cooling, amenity and ecosystem health—as demonstrated in this chapter by the modified design for the green park zone in the centre of Wielwijk, Dordrecht;
> The engagement process with the stakeholders makes it possible to take account of local values and sensibilities, which contributes to increased public and political support for the adaptive strategies, especially when the local environment will be significantly improved by the implementation of the options.

The ATP-AMO method has been successful in assessing the system robustness and adaptive potential of urban stormwater systems for flood risk and underscores the significance of project-level adaptation mainstreaming. The method illustrated can be readily applied to other key drivers and objectives and also to the integrated urban water cycle. A catalogue of critical ATPs could be defined, covering water quantity, quality, amenity and other criteria, limited only by the analytical capabilities and understandings covering the chemical, physical and social performance of these systems.

# 7. Comparing Real In Options and Adaptation Tipping Points

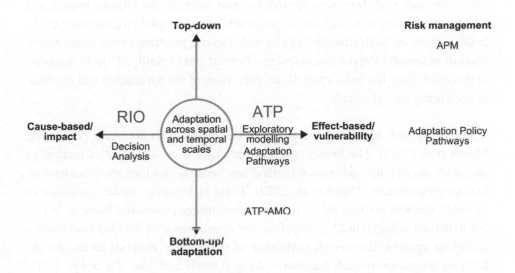

This chapter is based on:
Gersonius, B., Ashley, R., Jeuken, A., Pathirana, A. & Zevenbergen, C. 2012. Accounting for uncertainty and flexibility in flood risk management: comparing Real In Options and Adaptation Tipping Points. *Journal of Flood Risk Management*, under review.

## 7.1. Introduction

The planning/modification of flood risk management (FRM) systems needs to take account of climate change uncertainty. This is because of two features associated with such systems. First, the consequences of choices made as to the form and function of these systems have to be lived with for a long time, which means that the associated uncertainties are large. Second, potential irreversibilities in investment decisions can lead to a need for larger construction initially, particularly when using hard structural measures; which provides headroom for later adjustment. As an example: major sewers, once constructed, cannot be adapted or only with high cost. Improved approaches and methods for climate impact and adaptation assessment to address the uncertainties are needed (Kundzewicz et al., 2008); otherwise, such strategies can be maladaptive, resulting in e.g. unnecessary costs of potentially irreversible measures (Barnett and O'Neill, 2010). In response to this need, there has been a significant expansion of the approaches and methods in use during the last decade.

Different methods have been compared by Dessai and Van der Sluijs (2007) and Means et al. (2010). The former conclude that there is no 'silver bullet' method to deal with uncertainty; rather each method has benefits and limitations under particular circumstances (Carter et al., 2007). There is, however, limited guidance as to which methods are best for particular circumstances (Jones and Preston, 2011). An important research need is, therefore, the development of detailed case studies to further examine the overall usefulness of the various methods in use for informing adaptation-related decision making (Dessai and Van der Sluijs, 2007; Means et al., 2010). The review and comparison of methods within the resilience approach will be of particular significance, given the limited experience with this approach. This chapter contributes to this research need by discussing practical experience with two methods: Real-In-Options (RIO) and Adaptation Tipping Points (ATP). These methods both provide insight into and promote the ability of the system to deal with future change and, thus, can be used within the resilience approach. However, they have considerable differences in e.g. main orientations and application, and this is explained in the next sections.

## 7.2. Real-In-Options

### Concept in brief

A real option (RO) has been defined as 'the right—but not the obligation' to adjust the technical/infrastructure system in ways likely to be more resilient, as needed to continue to function as required in the face of change. As such, these options represent physical choices about the system that provide the flexibility to deal with uncertainty. ROs can be categorised as those that are either 'on' or 'in' systems (De Neufville, 2003). ROs 'on' systems are options applied to the decision making process related to infrastructure investment, treating the system configuration as a black box. ROs 'in' systems (RIO) are created by changing the engineering system (re)design as uncertainty is reduced. In the RIO concept, the options relate to the technical characteristics of the system, which cannot be known if the system configuration is treated as a black box.

## Procedure in brief

As given in Sect. 4.3, the RIO optimisation procedure comprises of the following steps:

1. Specify the scenario tree for the stochastic process that the uncertain variable follows.
2. Identify the potential options, or flexibility, in the technical/infrastructure system. The flexibility is provided by the design variables that can be changed after the initial implementation of a configuration.
3. Formulate the RIO optimisation problem in terms of its objectives, constraints and decision variables.
4. Establish and run the RIO optimisation model.

## Benefits and limitations

RIO analysis has considerable potential to support adaptation-related decision making, particularly if the required adaptation is significantly sensitive to the magnitude of climate change. The benefit of this method is that it identifies the "optimal" set of static adaptive strategies in response to changes in knowledge about climate change. This will allow the size and timing of investments to be informed by new information regarding (observed) climate change, and so improve the economic efficiency of such investments. The total life cycle costs will therefore likely be lower, as demonstrated by recent applications in the context of FRM (e.g., Gersonius et al., 2010; Woodward et al., 2011). Using RIO analysis also reduces initial capital costs and spreads the costs over the system lifetime which is important, especially in a time of economic stringency.

The theoretical limitation of RIO analysis is that it assumes probabilities can be given to future sea levels, river flows and rainfalls under climate change; although many climate scientists do not believe this is yet possible. Practical limitations are that it is complicated to establish and time-intensive to run the RIO optimisation model (Means et al., 2010), due to the potentially large number of design variables, time periods and boundary conditions (Wang and De Neufville, 2004). In addition, there is no clear procedure for monitoring information regarding e.g.

climate change in order to update/reassess the "optimal" dynamic adaptive strategy (*ibid*). Yet, it can be argued that such a procedure is not necessary; once new climate change scenarios become available, it is obvious that a new RIO analysis has to be carried out, using the new scenarios. However, this will introduce a new challenge as the underlying calculations and optimisation runs have to be repeated every time new scenarios become available. While being technically feasible, these calculations will require extra financial and/or personnel resources. Secondly, the "optimal" dynamic adaptive strategy will also change as new scenarios become available. This could complicate the implementation of the adaptive strategy, because of e.g. administrative, operational or legal aspects.

## 7.3. Adaptation Tipping Points

### Concept in brief

ATPs are the physical boundary conditions where acceptable technical, environmental, societal or economic standards may be compromised (Haasnoot et al., 2011). It is of note that ATPs, defined in this way, are only dependent on the magnitude of climate (and other) change(s) and not on time. This makes them much less dependent on climate change scenarios. Once determined, such points can be positioned in time using climate change scenarios. Combining the defined ATPs with climate change scenarios will provide information about the system robustness/resilience to climate change and the potential need for alternative adaptive strategies. The analysis can, therefore, help to develop Adaptation Pathways. This refers to a sequence of responses and potential adaptation options, which may be triggered before an ATP occurs (*ibid*).

### Procedure in brief

As given in Sect. 6.2, the steps for the ATP method are:
1. Specify the functions and climate change effects of interest for the assessment. The objectives for these functions (which are often translated into acceptable standards) are also defined. In addition, the current strategy to achieve the objectives should be identified.
2. Quantify the particular threshold values for the acceptable standards. These threshold values can be defined either according to regulations (e.g., by national law) or decided by the stakeholders involved and can change over time.
3. Identify the ATPs by increasing the design loading (e.g., the rainfall intensity) on the system, as a function of time, to assess the specific boundary conditions (i.e., the magnitude of climate change) under which acceptable standards may be compromised.

4.  Transform the specific boundary conditions under which an ATP will occur into an estimate of when it is likely to occur, using climate change scenarios.
5.  If it is desired that an ATP should not be reached, identify the potential options for adapting the system.
6.  Repeat steps 3 and 4 for the potential options identified; this will result in the definition of a number of adaptive strategies, some structural and some non-structural. Engagement with all stakeholders is required to select an adaptive strategy that is realistic and acceptable.

## Benefits and limitations

The ATP method examines the effects of increasing design loadings on the system performance. The benefit of this method is that it is virtually independent of climate change scenarios, and in particular of probabilities of climate change. Climate change becomes relevant for adaptation-related decision making only if it would lead to the crossing of an acceptability threshold. The ATP method, therefore, requires a range of plausible scenarios that can be used to assess whether or not the system is likely to cross any acceptability threshold in the face of change. In this sense, the method is more dependent on stakeholder engagement to quantify the acceptability thresholds, to identify the potential options for adapting the system, and to select an adaptive strategy that is realistic and acceptable. Kwadijk et al. (2011) point out that the application of ATPs answers the basic question of decision makers and other stakeholders: How much climate change can the current strategy cope with? They found that expressing uncertainty in terms of the period for which the current strategy is effective (i.e., when an ATP will be reached), appears more understandable for decision makers, than defining the likelihood of a specific outcome in a specific time period. Based on these findings, they conclude that the ATP method is useful to reduce the complexity of effect-based approaches. Another benefit identified by Kwadijk et al. is that the ATP method can provide easier updates, when new climate change scenarios become available. In addition, Chapter 6 recognises that the ATP method allows easy integration with the bottom-up approach. This approach commences at the local scale, assessing the system to determine whether it is feasible to increase its ability to deal with climate change, including the variability (Jones and Boer, 2005). The integration of the ATP method with the bottom-up approach helps to increase the no/low-regret character of adaptation for several reasons: it focuses the attention on the most urgent effects of climate change; it is expected to lead to potential cost reductions, since adaptation options can be integrated into the infrastructure and building designs at an early stage instead of being applied separately; it will lead to the development of area-specific responses, which could not have been developed on a higher scale level; it makes it possible to take account of local values and sensibilities, which contributes to increased public and political support for the adaptive strategies.

Although the outputs from the ATP method are easier to understand for decision makers, it can still be difficult to make a decision based on the outputs. This is because the method forces the decision makers to decide explicitly, through their choice of the preferred adaptive strategy, on those future conditions under which the system is likely to lack resilience (Lempert et al., 2004). Furthermore, expressing uncertainties about the system reaction with ATPs leads to a pseudo-certainty effect. An ATP depicts a definite limit (i.e., the point of reference), which is a simplification of the actual system reaction. The ATP method does not represent uncertainties associated with, for example, model simplification and the choice of the acceptability threshold. The treatment of uncertainty is therefore incomplete, and there is an argument here to use other methods in parallel, such as Monte-Carlo simulation, to quantify the uncertainty about the system reaction. A further limitation of the ATP method is that it uses up all the overcapacity in the STS for a single driver (i.e., climate change). Yet, the overcapacity is typically intended for a range of uncertain drivers/pressures, not just for climate change. Finally, the sensitivity analysis over the range of future conditions, as needed to determine the specific boundary conditions under which the acceptable standards may be compromised, can be a time-intensive process due to the potentially lengthy run times of hydrological/hydraulic models. The latter limitation could, potentially, be overcome by replacing complex detailed models with "low-resolution models" (Davis and Bigelow, 1998) or "Fast Simple Models" (Van Grol et al., 2006), as demonstrated by Haasnoot et al. (2012).

## 7.4. Choosing a method

Table 7.1 summarizes the two methods selected in terms of the following characteristics:

1.  Approach to cause and effect chain: identifies the direction in which the cause and effect chain (e.g., from pressure to state to impact) is followed in the reasoning;
2.  Ease of integration with the bottom-up approach: defines the ease in combining the method with the bottom-up approach (i.e. focusing on the local scale);
3.  Scenario requirements: considers the type of scenarios needed to apply the method;
4.  Engagement of decision makers: defines the need for input required from decision makers and other stakeholders for implementation and subsequent decision making;

5. Ease of understanding: defines the ease in grasping the concept and procedure of the method;
6. Ease of communication: defines the ease in communicating the results of the method to decision makers and other stakeholders;
7. Ease of development: considers the capacities and capabilities needed to develop the method;
8. Ease of implementation: considers the level of effort needed to implement the method;
9. Ease of output use: defines the ease in interpreting and making decisions based on the results of the method;
10. Ease of updating: defines the ease in updating/reassessing the adaptive strategy selected, when new climate change scenarios become available.

The characteristics in Table 7.1 have been modified from Means et al. (2010) where appropriate, and evaluated in two case studies presented in this thesis, namely West Garforth (England) and Wielwijk, Dordrecht (the Netherlands). For reasons of clarity not all characteristics of Means et al. (2010) have been reproduced in Table 7.1, but only those most important to assist in choosing a method. In addition, two more characteristics have been added that provide crucial information about the methods: the approach to cause and effect chain and the ease of integration with the bottom-up approach. As an example: the case study applications showed that the ease in combining the method with the bottom-up approach will be particularly significant where local stakeholders (including the public) are involved in decision making processes, such as in England where the localism agenda is now encouraging this (UK Parliament, 2010).

These characteristics in Table 7.1 can be used as a starting point for identifying which method to use under what circumstances. The selection of a method will depend on a number of factors, including: (1) knowledge about the probabilities of climate change; (2) agreement on the potential options for adapting the system; and (3) the capacities and capabilities available on the part of the user(s) of the method. If probabilistic scenarios are available and there is sufficient agreement on the potential options for adapting the system, then RIO analysis may be most appropriate as an attempt to identify an "optimal" dynamic adaptive strategy; as suggested by Defra (2009). This will, however, depend on the capacities and capabilities available to effectively use the probabilistic scenarios in optimisation procedures and, where integration with the bottom-up approach is envisaged, on the capacities of local stakeholders to engage in the decision making process. If probabilities of climate change are not available and/or there is not yet agreement on the potential options for adapting the system, then the ATP method may be most appropriate in an attempt to quantify and communicate the risks and identify a socially acceptable adaptive strategy. This is because the method is one which can be easily understood and used by decision makers and others, and is more accessible to local stakeholders. This will be dependent on the capacities of local stakeholders to engage in the decision making process.

**Table 7.1.** Key characteristics of each method (adapted from Means et al. 2010)

| Characteristic | Real In Options | Adaptation Tipping Points |
|---|---|---|
| Approach to cause and effect chain | Cause → Effect | Effect → Cause |
| Ease of integration with the bottom-up approach | Medium – It is more difficult to combine the method with the bottom-up approach, as there is less engagement in decision making | High – The method allows easy integration with the bottom-up approach |
| Scenario requirements | Probabilistic scenarios | Plausible scenarios |
| Engagement of decision makers | Medium – Engagement in initial steps to select acceptable standards and to identify potential options for adapting the system | High – Engagement throughout to give input on acceptable standards, potential options for adapting the system and the preferred adaptive strategy |
| Ease of understanding | Low – Concept of flexibility and procedure are complex to grasp, as are GAs | High – Concept of ATPs and procedure are easy to grasp |
| Ease of communication | High – The dynamic adaptive strategy can be easily explained and represented in a scenario tree | High – Simple in illustration when engaging with decision makers and other stakeholders |
| Ease of output use | High – The RIO model identifies the optimal dynamic adaptive strategy to climate change | Medium – Due to trade-offs between resilience and adaptation cost, it can be difficult to make a decision based on the outputs |
| Ease of development | Low – Establishing the RIO model is complicated and requires special skills, such as mathematical programming and computer programming | High – Developing the sensitivity analysis over the range of future climate conditions is straightforward |
| Ease of implementation | Medium – Running the RIO optimisation model can be time-intensive due to potentially large number of generations for the GA as well as the potentially lengthy run times of hydrological and/or hydraulic models | Medium – Executing the sensitivity analysis over the range of future climate conditions can be time-intensive due to the potentially lengthy run times of hydrological and/or hydraulic models |
| Ease of updating | Low – There is no clear procedure for monitoring information in order to update/reassess the managed/adaptive strategy selected | High – The method allows easy updating, when new climate change scenarios become available |

## 7.5. Discussion and conclusions

Two methods have been reviewed and compared that can be applied within the resilience approach: RIO and ATP. It has been concluded that each method has particular benefits under particular circumstances. It is, furthermore, of note that the methods are not mutually exclusive. In complex cases, such as river systems, it could be useful to apply the two methods in conjunction. As an example of this: ATP might be adopted at the start of the analysis process to short-list alternative adaptive strategies. The short-listed strategies are then analysed with the help of RIO, so as to optimise the size and timing of the adaptation options identified as part of a particular adaptive strategy. In the subsequent decision making process, the economic efficiency of the alternative adaptive strategies is compared in order to select the preferred adaptive strategy.

The fact that RIO and ATP use different approaches to cause and effect chain suggests that they can be incorporated into an overarching framework or process for facilitating resilience-focused adaptation; and this has been identified as an important research need (Carter et al., 2007). Rahman et al. (2008) identified Adaptive Policy Making (APM) (Walker, 2001) as an appropriate overarching framework. This provides a systematic method for monitoring the performance of a strategy, gathering information, implementing responses and potential adaptations over time, and adjusting and readjusting to new conditions. In the APM method, the strategies themselves would be designed to be incremental, adaptive, and conditional.

# 8. Conclusions and recommendations

## 8.1. Introduction

This chapter combines the results of this thesis to answer the research questions that, in Chapter 1, were derived from the general objective. The conclusions are split into conclusions regarding the sub questions and the main conclusion. Recommendations for practice and further scientific research for advancing resilience-focused adaptation are also given in this chapter.

## 8.2. Conclusions

### Conclusions regarding the sub questions

1.  How can resilience, and closely related terms, be defined be defined and assessed for STS?

Resilience has been used extensively in many studies on social-ecological systems (SES), but rather fewer studies have applied it to STS. The application of resilience with respect to STS differs from social-ecological studies in a number of respects, most notably the object (i.e., the resilience "of what") and the threshold type being considered in the assessment.

For SES, structural resilience has been treated as synonymous with functional resilience. For STS, however, it is crucial to make a distinction between the resilience of structures and functions. This is because resilient individual structures at particular scales (e.g., large-scale engineering structures or tightly regulated institutions) can, in some cases, threaten the performance of the function provided by the system. The aim of resilience management is, therefore, to enhance or maintain the performance of the function of interest and also to preserve those structures (both technical and social) that lead to enhanced or the same performance—and not necessarily preserve the existing systems themselves (Smith and Stirling, 2010).

The resilience of SES has been assessed by the distance of the system from the limit of the attraction basin, which is an ecological threshold. Many (but not all) ecological thresholds are non-returnable thresholds. STS do not exhibit ecological thresholds, but acceptability thresholds. These set out the requirements of system performance. The crossing of acceptability thresholds does not lead to irreversible system changes: it simply means that that a change in the adaptive strategy is required to restore the system performance to its original identity.

It is contended in Chapter 2 that required performance is best conceptualised in terms of maintaining identity in order to ensure that the STS has adequate resilience. Identity comprises the four aspects (components, relationships, innovation, and continuity) that constitute the minimum of what has to be identified and specified if resilience is to be assessed. Characterising the identity of a STS requires the conceptualisation of these four aspects in relation to the particular function provided by the system (e.g., flood risk management) and also the identification of the specific variables and thresholds that reflect changes in identity. A critical identity threshold occurs e.g. where the system performance is outside the acceptable standard. Using the identity approach, socio-technical resilience has been defined in Chapter 2 as "the ability of the system to continue to function as required in the face of change". This definition implies that a STS is resilient when it can deliver performance over the assessment period without a change of identity by continuing compliance with standards and expectations. If as a result of future change, the STS can no longer deliver required performance, then it may be considered as a different system: i.e., it changes its identity. The magnitude of change beyond which the system identity changes then becomes a fixed point of reference against which the degree of resilience can be quantified.

The above definition of socio-technical resilience might be considered ambiguous, because it makes no clear distinction between the various ways in which a STS can maintain its identity. Further clarification is, therefore, provided here. As discussed in Sect. 1.4, social-ecological resilience emerges from three complementary aspects: resilience as persistence, adaptability and transformability. These aspects are equally relevant for STS, and this can be summarised as follows.

Resilience as persistence refers as the ability of the technical/infrastructure system to absorb change, so as to continue to function as required in the face of change. This has been defined as technical/infrastructure resilience (Defra, 2011) or system robustness (Anderies et al., 2004). The boundary of technical/infrastructure resilience can be determined by assessing the occurrence of the critical ATP with respect to the existing technical/infrastructure system (e.g., the flood risk infrastructure).

Socio-technical resilience is a much broader concept than technical/infrastructure resilience: it involves the dynamic interplay between resilience as persistence, adaptability and transformability. Because the STS (as a whole) is dynamic, the ATPs have to be identified not only for the existing technical/infrastructure system, but also for the evolved system: i.e., after incremental adjustments have been implemented by the actors. The potential boundaries of socio-technical resilience can be determined by assessing the occurrence of the critical ATPs with respect to the existing STS (as a whole), with the current adaptive strategy. This approach for assessing socio-technical resilience recognises that, in order to continue to function as required in the face of change, incremental adjustments will logically be triggered in the existing system before ATPs occur.

In some cases, however, incremental adjustments will be too expensive or ineffective to maintain required performance; and a transformational change may be required. A transformational change refers to the implementation of a different kind of STS through a revision of the adaptive strategy. Hence, it involves 'breaking up' the structural resilience of the existing system in order to maintain or enhance functional resilience under changing future conditions. Socio-technical studies sometimes distinguish between two kinds of change processes, i.e. a transformation and a transition. A transformation is a change that is implemented by the regime actors, and a transition is a change that is driven by outside actors that develop radically new innovations (Geels and Kemp, 2007).

The most significant terms above are illustrated in Fig 8.1 for a hypothetical FRM system under the influence of climate change. The system comprises a low lying delta protected by dikes and a storm surge barrier. The boundary of technical/infrastructure resilience (or: system robustness) is indicated by the pink dot in Fig. 8.1. At this magnitude of sea level rise / river discharge, the elevation of dike section A will no longer meet the acceptable standard. Therefore, an intervention will be required to restore the system performance to its original identity, as shown in Fig 8.1. In strategy X (i.e., the current strategy), the existing FRM system is improved through dike heightening and replacing the existing barrier. This strategy involves incremental adjustments only. Strategy Y involves a transformation to a different kind of FRM system. This is achieved through the construction of new locks and re-arranging the river flows; these adaptation responses can be implemented by the regime actors (e.g., flood management organisations). Strategy Z, finally, involves a transition to a different kind of FRM system. It comprises transforming the existing dikes into overflowable dikes, together with sustainable spatial planning to reduce the consequences of flooding. Because spatial planning is the responsibility of the actors outside the existing regime (e.g., provinces and municipalities), a change of the regime actors is required in order to implement this strategy. The potential boundaries of socio-technical resilience of the three alternative strategies are indicated by the red dots in Fig. 8.1. The existing FRM system, with the current adaptive strategy, has a limited degree of resilience until 2100, because the boundary of resilience is likely to be crossed under both the medium and high climate change scenario for 2100. Strategy Y and strategy Z both have a high degree of resilience to climate change until 2100. The selection of the preferred strategy will involve comparing the potential degrees of resilience provided by each strategy with the associated costs and (wider) benefits.

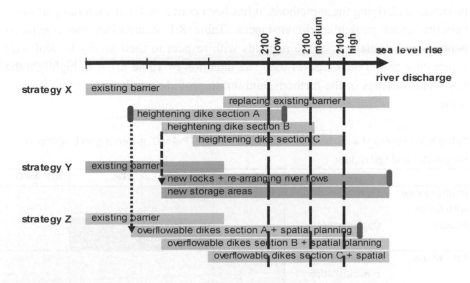

**Figure 8.1.** The occurrence of ATPs for the different strategies for adapting a hypothetical FRM system to climate change. To clarify: (i) the ends of the bar charts represent the ATPs; (ii) the purple block represents the boundary of technical/infrastructure resilience; (iii) the red blocks represent the potential boundaries of socio-technical resilience; (iv) the dashed line represents a transformation to a different kind of FRM system; (v) the dotted line represents a transition to a different kind of FRM system (adapted from: EA, 2009; Jeuken and Te Linde, 2011)

2. Which methods can be used within the resilience approach for climate impact and adaptation assessment? What are the benefits and limitations of the different methods?

The methods required to apply the resilience approach for climate impact and adaptation assessment should give insight into and promote the ability of the system to deal with future change. This comprises the ability not only to respond to threats (with in-built flexibility), but also to take advantage of opportunities that arise from future change. This thesis has provided (case study) experience with four methods within the resilience approach: Adaptive Policy Making (APM), Real-In-Options (RIO), Adaptation Tipping Point (ATP) and Adaptation Tipping Point - Adaptation Mainstreaming Opportunity (ATP-AMO). A larger range of methods can be applied within the resilience approach, but these were selected in order to cover a range of different approaches in combination with the resilience approach. APM combines the resilience approach with the risk management framework, RIO with the cause-based/impact approach, ATP with the effect-based/vulnerability approach and ATP-AMO with a hybrid approach that is both effect-based/vulnerability and bottom-up/adaptation. In light of the different approaches underlying these methods, it has been concluded that each has particular benefits under particular circumstances. Table 8.1 summarises the concluded benefits and limitations of the methods with respect to their ability to deal with future change and their ease-of-use. The intention of Table 8.1 is to highlight the key characteristics of the methods, and not to provide a fully comprehensive critique.

**Table 8.1.** Scorecard analysis of APM, RIO, ATP and ATP-AMO: green is good, yellow is moderate, and red is poor

| | | APM | RIO | ATP | ATP-AMO |
|---|---|---|---|---|---|
| **Ability to deal with future change** | Threats | | | | |
| | Opportunities | | | | |
| **Ease-of-use** | Interpretation and communication | | | | |
| | Development and implementation | | | | |
| | Decision making based on outputs | | | | |
| | Updating | | | | |

APM deals with change as a threat/opportunity by defining indicators and specific potential adaptations that can be taken in the future once certain thresholds or trigger events are reached. The main limitation of APM is the lack of a clear procedure for the development of a core strategy for maintaining required performance. Rather, it has broader utility as an overarching framework or process for facilitating resilience-focused adaptation. This framework is, therefore, best used in combination with other approaches to develop the core strategy. While the procedure of APM is relatively easy to understand, it can still be difficult to implement the method in practice. For example, the specification of the supporting strategy, which aims at addressing threats and opportunities, requires a detailed understanding of the wider contexts in which adaptation has to take place. As such, the method is better implemented within e.g. a Learning and Action Alliance context (e.g., Newman et al., 2011) than by experts alone.

The main benefit of RIO is its ability to deal with change as a threat by explicitly building in flexibility into the engineering system (re)design. This method uses probabilistic climate data to identify an "optimal" set of static adaptive strategies in response to advances in knowledge about climate change. This involves the estimation of the value of flexibility built into the engineering system (re)design. A major drawback is, however, that the method assumes probabilities can be given to future loadings under climate change; many climate scientists do not believe this is yet possible. Without probabilities the value of flexibility cannot be estimated. RIO does not provide a procedure for dealing with change as an opportunity, but could potentially be combined with the bottom-up approach to consider this ability. A practical benefit of RIO is that it is easy to make a decision based on the output of the analysis. This is because the method identifies which flexibilities are worth their cost. Limitations are that: the concept of flexibility and procedure are complex to grasp, as are GAs; the RIO optimisation model is complicated to establish and time-intensive to run; the "optimal" dynamic adaptive strategy will change as new scenarios become available.

ATP can deal with change as a threat by identifying and analysing potential options, or flexibility, for adapting the system to climate change. However, in its simplest form, it lacks a clear procedure for the development of an "optimal" dynamic adaptive strategy. In this respect, the recent extension of the ATP method, called Adaptation Pathways, provides a promising way forward—although this will come at the expense of the ease of development and implementation. ATP does not provide a procedure for dealing with change as an opportunity; though it can easily be combined with the bottom-up approach to consider this ability. The practical benefits of ATP are that: it is simple in concept and illustration when engaging with decision makers or other stakeholders; the analysis for this method is straightforward; it allows easy updating when new climate change scenarios become available. A limitation of ATP is that decision making can be difficult, because it forces decision makers to explicitly decide under which future conditions the system is likely to lack resilience.

ATP-AMO deals with change as a threat in exactly the same way as the ATP method. Its main benefit over ATP lies in its ability to deal with change as an opportunity. The ATP-AMO method provides a well-defined procedure for determining which responses and potential adaptations, where and when to incorporate into 'normal' investment projects, such as for urban regeneration and renewal. The implementation of ATP-AMO requires significant effort, due to the additional steps over the ATP method. The analysis of AMOs can be particularly challenging, since it requires the cooperation of different actors to share their investment agendas and bring these together in an attempt to save cost on adaptation and/or realise multi-beneficial opportunities at the same time. Because of the many other sectors involved (housing, traffic etc.), decision making on and updating the adaptive strategy is also more complex.

The selection of an appropriate method will depend on a number of factors, including (among others): knowledge about the probabilities of climate change; agreement on the potential options for adapting the system; and the capacities and capabilities available on the part of the user(s) of the method. As an example, Table 8.2 presents some recent case study experiences with the above methods, including the reasons for selecting the method.

**Table 8.2.** Recent experiences with the methods in practical cases

| Case study description | Method selected | Reason for selecting the method | Source |
|---|---|---|---|
| Water management in the Netherlands | ATP | It identifies the most urgent effects of climate change and accelerating sea level rise and when these effects will occur | Kwadijk et al. (2010) |
| Coastal management in the Netherlands | APM | It embeds the analysis process in an institutional framework; it considers different types of uncertainty | Rahman et al. (2008) |
| Defence raising for the Thamesmead area in the Thames Estuary | RIO | It identifies the "optimal" set of static adaptive strategies in response to advances in knowledge about climate and socio-economic change | Woodward et al. (2011) |
| Defence raising for the Rotterdam area in the Rhine-Meuse Estuary | ATP-AMO | It identifies where and when to incorporate defence raising with "normal" investment projects, such as for urban regeneration and renewal | Boer (2012) |

3.   What is the added value of the resilience approach for FRM?

FRM has evolved over time into its present form (e.g., Scrase and Sheate, 2005; Zevenbergen et al., 2012), and will continue to evolve in the future. This means that the added value of the resilience approach has to be understood in the light of the sequence of approaches used (through time) to deal with changing flood risk. An overview of these approaches is presented in Table 8.3, without attempting to be exhaustive. The details of the approaches have been discussed in Chapter 1, and are not repeated here.

**Table 8.3.** Sequence of approaches used (through time) to deal with changing flood risk

| Approach | Set of procedures | Assumption behind approach | Lesson learned about approach |
|---|---|---|---|
| Reactive approach | React to events that have occurred in the past | PDFs of future events cannot be known | Need to act on the likelihood that further events will occur in the future; PDFs of future events can be obtained through monitoring |
| Stationary hydrology approach | Monitor > Extrapolate > Act<br><br>*Note that this approach does not involve any adaptation, since the external drivers are assumed to be stationary* | External drivers are stationary; PDFs of future events will be the same as in the recent past | External drivers are non-stationary and system dynamics may compromise the stationarity assumption |
| Quasi-stationary hydrology approach | Monitor > Adjust for non-stationarity > Extrapolate > Adapt | PDFs of past events can be adjusted for non-stationarity through statistical analysis in order to obtain PDFs of future events | Climate change has introduced additional uncertainty, which makes adjustment through statistical analysis more difficult / impossible |
| Predict-Then-Adapt approach | Monitor > Predict > Adapt | Uncertainty about future loadings due to e.g. climate change can be reduced through prediction, allowing the adoption of a static robust or static adaptive strategy | Whilst climate science can potentially reduce part of the uncertainty about future loadings, there will always be significant irreducible uncertainty (e.g. related to future emissions) |
| Resilience approach | Iterative process of: Monitor & Predict & Adapt & Learn | Uncertainty reduction alone is not enough, and the uncertainty about future loadings has to be managed by adopting a dynamic adaptive strategy | *As of yet unknown due to limited practical experience with the approach* |

The added value of the resilience approach over the static approach derives from the understanding that the state of the FRM system is subject to change. This implies that the degree of system adaptedness to future conditions will change as the system context changes. Adaptedness refers to the effectiveness of the FRM system in meeting the requirements of performance in a specific system state. As an example: the FRM system may be initially designed to deal with the design loading under the medium climate change scenario (i.e., with a high degree of adaptedness), but it may be incapable of adapting to more extreme scenarios. Applying the methods within the resilience approach, e.g. RIO analysis, will provide insight into the trade-offs between the adaptedness of and the flexibility built into the FRM system. In this sense, the use of the resilience approach facilitates the development of responses and potential adaptations that are appropriate at the right time and right cost. The resilience approach, furthermore, suggests that future change, such as that which arises from urban dynamics, may create opportunities for adapting the FRM system to climate change. Application of the methods within the resilience approach, e.g. ATP-AMO, will help to identify and take advantage of the right opportunities.

### Main conclusion

Answering the three sub questions has contributed to the formulation of the main conclusion of this thesis, as stated below.

**The results of this thesis have shown that the resilience approach has significant potential to support the adaptation of FRM systems to climate change. Applying the methods within the resilience approach will provide insight into and promote the ability of the FRM system to deal with future change. This comprises the ability not only to respond to threats (with in-built flexibility), but also to take advantage of opportunities that arise from future change. It will, therefore, facilitate the development of responses and potential adaptations that are appropriate at the right time and right cost. It will also help to identify and take advantage of the right opportunities.**

## 8.3. Recommendations

### Recommendations for practice

1. Policy makers and practitioners will wish for certainty in the assessment of climate impacts and adaptation, but certainty cannot be guaranteed and therefore new and improved methods are needed to address the uncertainties and to help make the right decisions. The resilience approach facilitates the development of dynamic adaptive strategy in response to these uncertainties. It is, therefore, recommended to consider this approach in addition to the approaches given by the IPCC AR4: i.e., impact approach, vulnerability approach, adaptation approach and risk management framework (Carter et al., 2007).

2.  The shift toward the resilience approach will be particularly challenging to flood risk managers and engineers. For them the possession of good knowledge about future drivers (e.g., data about historical rainfall informing them about future conditions) and the effectiveness of responses (e.g., experience in designing and building flood risk infrastructure) has been fundamental to planning in the past. In this respect, there should be more concerted efforts to enhance the capacity in this group to start utilising and build experience with the new approaches and methods for climate impact and adaptation assessment.

3.  There needs to be more flexibility in risk standards and risk assessment for FRM systems. Current risk assessment is aimed at meeting pre-set acceptable standards, which are hardly or not at all debatable. In the Netherlands, for example, the acceptable standards for open water systems have been established by national law (NBW, 2003). These standards require the capacity of the open water system to be adequate in 2015 to deliver an acceptable risk under the medium climate change scenario for 2050. This fixed standard makes it more difficult for practitioners to take advantage of AMOs that are likely to occur beyond 2015, such as opportunities for incorporating adaptation responses into urban regeneration and renewal.

4.  There is a need to review and, where necessary, adjust institutional frameworks to facilitate resilience-focused adaptation. Institutional changes should include the support for closer interaction and communication between actor groups, for example, via Learning and Action Alliances (Termeer and Meijerink, 2008; Van Herk et al., 2011). Working in groups or alliances helps improve how multiple actors in the FRM system work (i.e., doing things differently) and can lead to interventions having greater impacts and/or lower costs. It also stimulates the different actors to open up their investment agendas and bring these together in an attempt to save cost on adaptation and/or realise multi-beneficial opportunities at the same time. There is, furthermore, a need (particularly) at the local level to provide updated mechanisms to secure funding for climate adaptation. As an example: regulations and acceptable standards for sewer systems in the Netherlands (e.g., Rioned, 2004) do not yet anticipate climate change. This means that there is no mechanism within the regulatory system to secure funding for building headroom and/or into flexibility the sewer system (i.e., with potential cost savings for later adjustment).

5.  There is a need to review and, where necessary, adjust the discount rate being used for FRM. The discount rate will significantly influence the economic value of being able to deal with future change as a threat and/or opportunity. This can be explained as follows. The economic value of being able to deal with change as a threat through flexibility reduces with higher discount rates.

This is because any cost savings by avoiding potentially unnecessary investments later will be discounted more heavily with higher discount rates. The economic value of being able to deal with change as an opportunity also reduces with higher discount rates. For example, it has been demonstrated that the attractiveness of project-level adaptation mainstreaming is significantly sensitive to the discount rate (see Table 6.1). The higher the discount rate the more attractive it becomes to delay potential adaptations until the occurrence of a critical ATP, instead bringing these adaptations forward in time to implement them synergistically with "normal" investment projects. In the light of the above, the use of a declining discount rate over time is recommended for FRM (over a constant discount rate of e.g. 4 % or more), because this places a higher economic value on the ability to deal with future change. As an example: in the Netherlands, a constant discount rate of either 4 or 5.5 % is used for FRM (Financiën, 2009), which significantly decreases the economic value of being able to deal with future change—yet, this ability is promoted as a shared value of the Dutch Delta Programme.

**Recommendations for further scientific research**

1. In this thesis, a STS is considered resilient when it can deliver performance over the assessment period without a change of identity by continuing compliance with standards and expectations. Because standards and expectations will change in the future, the quantitative threshold values used to define identity changes will also be emergent (i.e., identity is a dynamic property). It is, therefore, not possible to determine the potential boundaries of resilience with any certainty into the future. An important research need is to understand how social values and interests may evolve over time in response to socio-economic and/or climatic change and to consider this in relation to the system's identity (refer e.g. to Haasnoot et al. (2012) for an example of how this can be done). Understanding these change processes will be crucial for the assessment and management of socio-technical resilience.

2. The case study applications in this thesis all start from the notion of a given acceptable standard (e.g., based on the probability of flooding). The particular threshold value for the acceptable standard will strongly determine the results of the assessment of socio-technical resilience. Alternatively, the magnitude of the threshold value itself could be viewed as a factor which can be optimized with respect to other underlying objectives (e.g., economics). This alternative view and its implications should be given more attention in further scientific research. In FRM, this view is reflected in the risk-based approach, where the acceptable standard is the economically defined optimal level of flood risk in terms of costs and benefits.

3. This thesis has applied the resilience approach to the adaptation of FRM systems to climate change. It is recommended to also examine the usefulness of this resilience approach for other water systems, such as water supply and pollution control, and for (or rather: in combination with) other external drivers, such as population growth and economic development. It is of note that others have done this already, but using older resilience theories.

4. This thesis has provided experience with four methods within the resilience approach: Real In Options (RIO); Adaptation Tipping Points (ATP); and Mainstreaming. It is recommended to further study and evaluate other methods within the resilience approach, such as decision tree analysis (De Bruin and Ansink, 2011), exploratory modelling (Lempert et al., 2003), Adaptation Pathways (Haasnoot et al., 2012) and Adaptation Policy Pathways (Walker et al., 2012), and to gain experience with these methods in practical cases for FRM systems.

5. The focus of this thesis was mainly on promoting incremental adjustment to maintain or enhance the resilience of FRM systems to climate change. In some cases, however, incremental adjustments will be too expensive or ineffective to maintain required performance; and a transformational change may be required. More research is needed to understand the requirements for promoting transformational change in relation to (socio-technical) resilience.

# 9. Bibliography

Addis, B. 2007. Building: 3000 years of design engineering and construction. Phai-don Press, London, ISBN 978 0 7148 4146 5.

Agusdinata, D. 2008. Exploratory modeling and analysis. Next Generation Infrastructures Foundation, Delft, The Netherlands.

Amram, M. & Kulatilaka, N. 1999. *Real options: managing strategic investment in an uncertain world*, Boston, Mass.: Harvard Business School Press.

Anderies, J., Janssen, M. & Ostrom, E. 2004. A framework to analyze the robustness of social-ecological systems from an institutional perspective. *Ecology and Society,* 9, 18.

Antikarov, V. & Copeland, T. 2001. Real options: A practitioner's guide. *New York.*

Arkell, B., Darch, G., Jeal, G., Jones, P., Kilsby, C., McSweeney, R., Osborn, T. & Ravnkilde, K. 2010. Climate Change Modelling for Sewerage Networks UKWIR Report CL10. UK Water Industry Research, London, UK.

AS/NZS 4360 2004. Tutorial Notes: Broadleaf Capital International Pty Ltd, 2004.

Ashley, R., Blanksby, J. & Cashman, A. 2008. Adaptable Urban Drainage – Addressing Change in Intensity, Occurrence and Uncertainty of Stormwater (AUDACIOUS). A Main summary report – final draft 1.1.

Ashley, R., Blanskby, J., Newman, R., Gersonius, B., Poole, A., Lindley, G., Smith, S., Ogden, S. & Nowell, R. 2011. Learning and Action Alliances to build capacity for flood resilience. *Journal of Flood Risk Management.*

Ashley, R. & Brown, R. 2009. Entrapped in common sense: why water management by current regimes is not sustainable and what we can do about it. Proc. 9th NESS: Knowledge, learning and action for sustainability, 10th-12th June 2009, London.

Ashley R M., Nowell R., Gersonius B. & Walker L. 2011. Surface Water Management and Urban Green Infrastructure. Review of Current Knowledge. Foundation for Water Research FR/R0014 May. 73pp.

Barnett, J. & O'neill, S. 2010. Maladaptation. *Global Environmental Change,* 20, 211-213.

Berkes, F., Colding, J. & Folke, C. 2003. *Navigating social-ecological systems: building resilience for complexity and change*, Cambridge Univ Pr.

Berkes, F., Folke, C. & Colding, J. 2000. *Linking social and ecological systems: management practices and social mechanisms for building resilience*, Cambridge Univ Pr.

Black, F. & Scholes, M. 1973. The pricing of options and corporate liabilities. *Journal of Political Economy,* 81, 637.

Boer, F. 2012. Two-pager sub areas Rhine Meuse Delta, IABR 2012. Two-pager deelgebieden Rijn-Maasdelta, IABR 2012. (In Dutch). Unpublished paper.

Brand, F. 2009. *Resilience and Sustainable Development: An Ecological Inquiry.*

Brand, F. S. & Jax, K. 2007. Focusing the Meaning (s) of Resilience: Resilience as a Descriptive Concept and a Boundary Object. *Ecology and Society,* 12, 23.

Brooks, N. & Adger, W. N. 2005. Assessing and enhancing adaptive capacity. *2005): Adaptation Policy Frameworks for Climate Change: Developing Strategies, Policies and Measures, Cambridge University Press, Cambridge, UK, S,* 165-181.

Brown, R. R., Ashley R. M. & Farrelly, M. 2011. Political and Professional Agency Entrapment: An Agenda for Urban Water Research. Water Resources Management. Vol. 23, No.4. European Water Resources Association (EWRA). ISSN 0920-4741. DOI 10.1007/s11269-011-9886-y

Brugnach, M., Dewulf, A., Pahl-Wostl, C. & Taillieu, T. 2008. Toward a relational concept of uncertainty: about knowing too little, knowing too differently, and accepting not to know. *Ecology and Society,* 13, 30.

BSI 1995-1998. BS EN 752, Drain and Sewer Systems Outside Buildings, Parts 1-7.

Buishand, A. & Wijngaard, J. 2007. Statistiek van extreme neerslag voor korte neerslagduren, KNMI rapport TR-295, De Bilt.

Burton, I. & May, E. 2004. The adaptation deficit in water resource management. *IDS bulletin,* 35, 31-37.

Buurman, J., Zhang, S. & Babovic, V. 2009. Reducing Risk Through Real Options in Systems Design: The Case of Architecting a Maritime Domain Protection System. *Risk Analysis,* 29, 366-379.

Carpenter, S., Walker, B., Anderies, J. M. & Abel, N. 2001. From Metaphor to Measurement: Resilience of What to What? *Ecosystems,* 4, 765-781.

Carter, T. R., Jones, R., Lu, X., Bhadwal, S., Conde, C., Mearns, L., O'neill, B., Rounsevell, M. & Zurek, M. 2007. New assessment methods and the characterisation of future conditions. *Climate Change 2007: Impacts, Adaptation and Vulnerability. Contribution of Working Group II to the Fourth Assessment Report of the Intergovernmental Panel on Climate Change, Parry, M.L., Canziani, O.F., Palutikof, J.P., van der Linden, P.J., and Hanson, C.E., Eds., Cambridge University Press, Cambridge, UK,* 133-171.

Chow, V. T., Maidment, D. R. & Mays, L. W. 2005. *Applied hydrology,* McGraw-Hill New York (NY).

Committee on Hydrologic Science 2011. *Global Change and Extreme Hydrology: Testing Conventional Wisdom,* Natl Academy Pr.

Cooper, D. 2004. The Australian and New Zealand Standard on Risk Management, AS/NZS 4360: 2004. *Tutorial Notes: Broadleaf Capital International Pty Ltd.*

Cox, J., Ross, S. & Rubinstein, M. 2002. Option pricing: a simplified approach'. *International Library of Critical Writings in Economics,* 143, 461-495.

Cox, P. & Stephenson, D. 2007. A changing climate for prediction. *Science,* 317, 207.

Cumming, G., Barnes, G., Perz, S., Schmink, M., Sieving, K., Southworth, J., Binford, M., Holt, R., Stickler, C. & Van Holt, T. 2005. An exploratory framework for the empirical measurement of resilience. *Ecosystems,* 8, 975-987.

Davis, P. K. & Bigelow, J. H. 2003. *Motivated metamodels: synthesis of cause-effect reasoning and statistical metamodeling,* Rand Corporation.

De Bruijn, K. M. 2004. Resilience indicators for flood risk management systems of lowland rivers. *International Journal of River Basin Management,* 2, 199-210.

De Bruin, K. & Ansink, E. (2011). Investment in flood protection measures under climate change uncertainty. *In:* European Association of Environmental and Resource Economists, 18th Annual Conference, 29 June - 2 July 2011, Rome

De Neufville, R. 1990. Applied systems analysis: engineering planning and technology management.

De Neufville, R. 2003. Real Options: Dealing With Uncertainty in Systems Planning and Design. *Integrated Assessment,* 4, 26-34.

De Neufville, R. 2004. Uncertainty Management for Engineering Systems Planning and Design. *Engineering Systems Symposium, MIT, Cambridge, MA.*

Deb, K., Pratap, A., Agarwal, S., Meyarivan, T., Fast, A. & Algorithm, E. 2002. NSGA-II. *IEEE transactions on evolutionary computation,* 6.

Defra 2006. Flood and Coastal Defence Appraisal Guidance FCDPAG3 Economic Appraisal. Supplementary note to operating authorities – climate change impacts.

Defra 2011. *Climate Resilient Infrastructure: Preparing Infrastructure for A Changing Climate,* The Stationery Office.

Defra & EA 2011. Understanding the risks, empowering communities, building resilience. The national flood and coastal erosion risk management strategy for England. Available at: www.official-documents.gov.uk.

Delta Committee 1960. Report of the Delta Committee. *Ministry of Water Works, The Hague, The Netherlands.*

Dessai, S., Hulme, M., Lempert, R. & Pielke Jr, R. 2008. Climate prediction: a limit to adaptation. *Living with climate change: are there limits to adaptation,* 49-57.

Dessai, S., Hulme, M., Lempert, R. & Pielke Jr, R. 2009. Do we need better predictions to adapt to a changing climate. *Eos Transactions AGU,* 90, 111-112.

Dessai, S. & Van Der Sluijs, J. 2007. Uncertainty and climate change adaptation: a scoping study. *Report NWS-E,* 1-95.

Dhondia, J. & Stelling, G. Year. Application of one-dimensional-two-dimensional integrated hydraulic model for flood simulation and damage assessment. *In:*

Proc. 5th Hydroinformatics Conference, Cardiff, United Kingdom, 2002. 265-276.

Dixit, A. K. & Pindyck, R. S. 1994. *Investment under uncertainty*, Princeton University Press Princeton, NJ.

EA 2009. Thames estuary 2100, managing flood risk through london and the thames estuary. Tech. rep., Environment Agency, Available at: http://www.environment-agency.gov.uk/research/library/consultations/06100 .aspx

EA 2012. Accounting for adaptive capacity in Flood and Coastal erosion risk management. Workshop, London. 1st March 2012.

European Commission 2009. White Paper on Adapting to climate change: Towards a European framework for action.

Evans, E. P., Ashley, R., Hall, J. W., Penning-Rowsell, E. C., Saul, A., Sayers, P. B., Thorne, C. R. & Watkinson, A. 2004. Foresight Flood and Coastal Defence Project: Scientific Summary: Volume I, Future Risks and Their Drivers. *London: Office of Science and Technology*.

Financiën 2009. Advisory Group Long Term Discount Rate. Advies Werkgroep Lange Termijn Discontovoet. (In Dutch).

Folke, C. 2006. Resilience: The emergence of a perspective for social–ecological systems analyses. *Global Environmental Change*, 16, 253-267.

Folke, C., Carpenter, S. R., Walker, B., Scheffer, M., Chapin, T. & Rockström, J. 2010. Resilience thinking: integrating resilience, adaptability and transformability. *Ecology and Society*, 15, 20.

Franssen, R., Leeuwen van, E. & Oostrom van, N. 2011. Climate adaptation for regeneration and the potential of tipping points. Klimaatadaptatie bij herstructurering en de potentie van knikpunten. (In Dutch). Deltares rapport 1203987-000.

Gallopín, G. 2006. Linkages between vulnerability, resilience, and adaptive capacity. *Global Environmental Change*, 16, 293-303.

Geels, F. W. 2004. From sectoral systems of innovation to socio-technical systems:: Insights about dynamics and change from sociology and institutional theory. *Research Policy*, 33, 897-920.

Geels, F. W. 2005. *Technological transitions and system innovations: a co-evolutionary and socio-technical analysis*, Edward Elgar Publishing.

Geels, F. W. & Kemp, R. 2007. Dynamics in socio-technical systems: Typology of change processes and contrasting case studies. *Technology in Society*, 29, 441-455.

Gersonius, B., Ashley, R., Pathirana, A. & Zevenbergen, C. 2010. Managing the flooding system's resiliency to climate change. *Proceedings of the ICE-Engineering Sustainability*, 163, 15-23.

Gersonius, B., Zevenbergen, C., Puyan, N. & Billah, M. 2008. Efficiency of private flood proofing of new buildings–Adapted redevelopment of a floodplain in the Netherlands. *WIT Transactions on Ecology and the Environment,* 118, 247-259.

Goldstein, J. 1999. Emergence as a construct: History and issues. *Journal of Complexity Issues in Organizations and Management,* 1.

Gunderson, L. & Holling, C. 2002. *Panarchy Synopsis: Understanding Transformations in Human and Natural Systems*, Island Pr.

HM Treasury 2003. *The green book: appraisal and evaluation in central government*, Stationery Office.

Haasnoot, M., Middelkoop, H., Van Beek, E. & Van Deursen, W. 2011. A method to develop sustainable water management strategies for an uncertain future. *Sustainable Development,* 19, 369-381.

Haasnoot, M., Middelkoop, H., Offermans, A., van Beek, E. & van Deursen, W.P.A. 2012. Exploring pathways for sustainable water management in river deltas in a changing environment. *Climatic Change.* DOI 10.1007/s10584-012-0444-2.

Haasnoot, M., Kwakkel, J.H. & Walker W.E. 2012. Designing Adaptive Policy Pathways for Sustainable Water Management under Uncertainty: Lessons Learned from Two Cases. *In:* Third International Engineering Systems Symposium CESUN 2012, Delft University of Technology, 18-20 June 2012.

Hall, J. W., Meadowcroft, I. C., Sayers, P. B. & Bramley, M. E. 2003. Integrated flood risk management in England and Wales. *Natural Hazards Review,* 4, 126.

Hawkins, E. & Sutton, R. 2009. The potential to narrow uncertainty in regional climate predictions. *Bulletin of the American Meteorological Society,* 90, 1095-1107.

HM Government 2011. Climate Resilient Infrastructure: Preparing for a Changing Climate. Available at: http://www.defra.gov.uk/publications/files/climate-resilient-infrastructure-full.pdf

HM Treasury & Defra 2009. Accounting for the Effects of Climate Change. June 2009 - Supplementary Green Book Guidance.

Holling, C. 1996. Engineering resilience versus ecological resilience. *Engineering within ecological constraints*, 31-43.

Holling, C. S. 1973. Resilience and Stability of Ecological Systems. *Annual Reviews in Ecology and Systematics,* 4, 1-23.

Hulme, M., Pielke, R. & Dessai, S. 2009. Keeping prediction in perspective. *Nature Reports Climate Change.*

Huq, S. & Reid, H. 2004. Mainstreaming adaptation in development. *IDS bulletin,* 35, 15-21.

IenM 2011. Main report of the 2nd Flood Protection Programme. Basisrapportage van het 2e Hoogwaterbeschermingsprogramma (HWBP-2). (In Dutch).

Ingham, A., Ma, J. & Ulph, A. 2006. Theory and Practice of Economic Analysis of Adaptation. Tyndall Centre for Climate Change Research, university of East Anglia.

Jarrow, R. & Rudd, A. 1982. Approximate option valuation for arbitrary stochastic processes. *Journal of Financial Economics,* 10, 347-369.

Jeuken, A. & Te Linde, A. 2011. Applying tipping points and adaptation pathways. Werken met knikpunten en adaptatiepaden. (In Dutch). Available at: kennisonline.deltares.nl

Jones, R. & Boer, R. 2005. Assessing current climate risks. *Adaptation Policy Framework for Climate Change: Developing Strategies, Policies and Measures, Ian Burton, Elizabeth L. Malone and Saleemul Huq (lead authors). Cambridge University Press, Cambridge.*

Jones, R. N. & Preston, B. L. 2011. Adaptation and risk management. *Wiley Interdisciplinary Reviews: Climate Change,* 2, 296-308.

Kabat, P., Van Vierssen, W., Veraart, J., Vellinga, P. & Aerts, J. 2005. Climate proofing the Netherlands. *Nature,* 438, 283-284.

Kundzewicz, Z., Mata, L., Arnell, N., Doll, P., Jimenez, B., Miller, K., Oki, T., Sen, Z. & Shiklomanov, I. 2008. The implications of projected climate change for freshwater resources and their management. *Hydrological Sciences Journal/Journal des Sciences Hydrologiques,* 53, 3-10.

Kwadijk, J. C. J., Haasnoot, M., Mulder, J. P. M., Hoogvliet, M., Jeuken, A., Van Der Krogt, R. A. A., Van Oostrom, N. G. C., Schelfhout, H. A., Van Velzen, E. H. & Van Waveren, H. 2010. Using adaptation tipping points to prepare for climate change and sea level rise: a case study in the Netherlands. *Wiley Interdisciplinary Reviews: Climate Change,* 1, 729-740.

Kwakkel, J., Walker, W. & Marchau, V. 2010. Adaptive airport strategic planning. *EJTIR,* 3.

Langston, C., Wong, F., Hui, E. & Shen, L.Y. 2008. Strategic Assessment of Building Adaptive Reuse Opportunities in Hong Kong, Building and Environment, 43(10), pp.1709-1718.

Larsen, A. N., Gregersen, I. B., Christensen, O., Linde, J. J. & Mikkelsen, P. S. 2009. Potential future increase in extreme one-hour precipitation events over Europe due to climate change. *Water science and technology,* 60, 2205-2216.

LCC, City of Bradford Metropolitan District Council, Yorkshire Water Services, Environment Agency & Pennine Water Group 2008. West Garforth Integrated Urban Drainage Pilot Study Final Report. Version 1.9 Final.

Lempert, R., Nakicenovic, N., Sarewitz, D. & Schlesinger, M. 2004. Characterizing Climate-Change Uncertainties for Decision-Makers. An Editorial Essay. *Climatic Change,* 65, 1-9.

Lempert, R. J., Popper, S. W. & Bankes, S. C. 2003. *Shaping the next one hundred years: new methods for quantitative, long-term policy analysis*, Rand Corp.

Lempert, R. J. & Schlesinger, M. E. 2000. Robust strategies for abating climate change. *Climatic Change,* 45, 387-401.

Lim, B., Spanger-Siegfried, E., Burton, I., Malone, E. L. & Huq, S. 2005. *Adaptation policy frameworks for climate change: developing strategies, policies and measures*, Cambridge University Press, Cambridge, UK.

Lowe, J., Howard, T., Pardaens, A., Tinker, J., Holt, J., Wakelin, S., Milne, G., Leake, J., Wolf, J. & Horsburgh, K. 2009. UK Climate Projections science report: Marine and coastal projections.

Luijtelaar, H. van, Luijendijk, J. & Lee, A. van der. 2006. Optimisation water system Dordrecht. Optimalisatie watersysteem Dordrecht. (In Dutch).

Marchaut, V., Walker, W. & Van Wee, B. Year. Innovative long-term transport policymaking: from predict and act to monitor and adapt. *In:* European Transport Conference Proceedings, 2007.

Means, E., Laugier, M., Daw, J., Kaatz, L. & Waage, M. 2010. Decision Support Planning Methods: Incorporating Climate Change Uncertainties Into Water Planning. *Water Utility Climate Alliance, San Francisco*.

Mens, M. J. P., Klijn, F., De Bruijn, K. M. & Van Beek, E. 2011. The meaning of system robustness for flood risk management. *Environmental Science & Policy*.

Merton, R. C. 1973. Theory of rational option pricing. *Bell Journal of Economics and Management Science,* 4, 141-183.

MfE 2010. Preparing for future flooding: A guide for local government in New Zealand. Wellington, 34 p.

Milly, P. C. D., Betancourt, J., Falkenmark, M., Hirsch, R. M., Kundzewicz, Z. W., Lettenmaier, D. P. & Stouffer, R. J. 2008. Climate Change: Stationarity Is Dead: Whither Water Management? *Science,* 319, 573.

Mitchell, G. 2001. The Quality of Urban Stormwater in Britain and Europe: Database and Recommended Values for Strategic Planning Models. Technical Report, School of Geography, University of Leeds.

Morselt, T.T. 2009. Economic and environmental effects of the sand extraction strategy. Economische en milieukundige effecten van de zandwinstrategie. (In Dutch). Rapport Blueconomy, P09014.

Murphy, J., Sexton, D., Jenkins, G., Booth, B., Brown, C., Clark, R., Collins, M., Harris, G., Kendon, E. & Betts, R. 2009. UK Climate Projections Science Report: Climate Change Projections. *Met Office Hadley Centre, Exeter, UK*.

Myers, S. 1984. Finance Theory and Financial Strategy. *Interfaces,* 14, 126.

NBW 2003. Publication The Netherlands is living with water. Publicatie Nederland leeft met water. (In Dutch).

Nelson, D. R., Adger, W. N. & Brown, K. 2007. Adaptation to environmental change: contributions of a resilience framework. *Annu. Rev. Environ. Resour.*, 32, 395-419.

New, M., Lopez, A., Dessai, S. & Wilby, R. 2007. Challenges in using probabilistic climate change information for impact assessments: an example from the water sector. *Philosophical Transactions A*, 365, 2117.

Newman, R., Ashley, R., Molyneux-Hodgson, S. & Cashman, A. 2011. Managing water as a socio-technical system: the shift from 'experts' to 'alliances'. *Proceedings of the ICE - Engineering Sustainability*, 164, 95 -102.

OFWAT. 2008. Preparing for the future – Ofwat's climate change policy statement. Ofwat, Birmingham. Available at: www.ofwat.gov.uk/sustainability/climate-change/pap_pos_climatechange.pdf

Olsen, J. R., Kiang, J. E., Waskom, R. M. T., Engineers, U. S. A. C. O., Institute, C. W. & University, C. S. 2010. *Workshop on Nonstationarity, Hydrologic Frequency Analysis, and Water Management: January 13-15, 2010, Boulder, Colorado*, Colorado State University, Colorado Water Institute.

Ottens, M., Franssen, M., Kroes, P. & Van De Poel, I. 2006. Modelling infrastructures as socio-technical systems. *International journal of critical infrastructures*, 2, 133-145.

Parry, M. & Carter, T. 1998. Climate impact and adaptation assessment: a guide to the IPCC approach.

Persson, Å. & Klein, R. 2009. Mainstreaming adaptation to climate change into official development assistance: Building on environmental policy integration theory. *Environmental Change and Foreign Policy: Theory and Practice.*

PZH 2011. Third Round Statutory Assessment Primary Flood Defences, Dike Ring Area 22 - Island of Dordrecht. Derde Ronde Toetsing Primaire Waterkeringen, Dijkringgebied 22 - Eiland van Dordrecht. (In Dutch).

Rahman, S., Walker, W. & Marchau, V. 2008. Coping with Uncertainties about Climate Change in Infrastructure Planing – An Adaptive Policymaking Approach, Final Report.

Rioned 2004. Urban Drainage Guideline, C2100 Hydraulic performance sewerage calculations. Leidraad Riolering, operationeel beheer, C2100 Rioleringsberekeningen, hydraulisch functioneren. (In Dutch). Kluwer, Alphen aan den Rijn, the Netherlands.

Rioned 2006. Urban water task, comparison standards for water on streets and inundation. Stedelijke wateropgave, vergelijking normen voor water op straat en inundatie. (In Dutch).

Rip, A. & Kemp, R. 1998. Technological Change. In: Rayner S., Malone EL (editors).

Rossman, L. 2004. Storm water management model (SWMM version 5.0) user's manual. *United State Environment Protection Agency.*

Scheffer, M., Carpenter, S., Foley, J. A., Folke, C. & Walker, B. 2001. Catastrophic shifts in ecosystems. *Nature,* 413, 591-596.

Scheffer, M. & Carpenter, S. R. 2003. Catastrophic regime shifts in ecosystems: linking theory to observation. *Trends in Ecology & Evolution,* 18, 648-656.

Schneider, S. 1983. CO2, climate and society: a brief overview. *Social Science Research and Climate Change: In Interdisciplinary Appraisal, D. Reidel, Boston,* 9-15.

Schultz, B. & Wandee, P. 2003. Some practical aspects of the new policy on water management in the Netherlands polders, International Institute for Land Reclamation and Improvement (ILRI).

Scrase, J.I. & Sheate, W.R. 2005. Re-framing flood control in England and Wales. *Environmental Values,* 14, 1, 113-137.

Shapiro, S. & Wilk, M. 1965. An analysis of variance test for normality (complete samples). *Biometrika,* 52, 591.

Smith, A. & Stirling, A. 2010. The politics of social-ecological resilience and sustainable socio-technical transitions. *Ecology and Society,* 15, 11.

Solomon, S., Qin, D., Manning, M., Alley, R., Berntsen, T., Bindoff, N., Chen, Z., Chidthaisong, A., Gregory, J. & Hegerl, G. 2007. Climate Change 2007: The Physical Science Basis, Contribution of Working Group 1 to the Fourth Assessment Report of the Intergovernmental Panel on Climate Change.

Swanson, D. & Bhadwal, S. 2010. *Creating Adaptive Policies: A Guide for Policymaking in an Uncertain World,* Sage Publications Pvt. Ltd.

Taylor, A., Wong, T. & Hydrology, C. R. C. F. C. 2002. *Non-structural stormwater quality best management practices: an overview of their use, value, cost and evaluation,* CRC for Catchment Hydrology.

TAW 2002. Technical Report Wave Run-up and Wave Overtopping at dikes.

Termeer, C. & Meijerink, S. 2009. Climate proof or climate neutral governance? An essay on the adaptive capacities of institutions. Klimaat bestendig of klimaat neutraal bestuur? Een essay over het adaptief vermogen van instituties. (In Dutch). Deeladvies en achtergrondstudie voor de Raad voor Verkeer en Waterstaat.

Thorne, C. R., Evans, E. P. & Penning-Rowsell, E. C. 2007. *Future flooding and coastal erosion risks,* Thomas Telford Services Ltd.

Trigeorgis, L. 1996. *Real options: Managerial flexibility and strategy in resource allocation,* the MIT Press.

UK Parliament. 2010. Localism Bill 2010-11. Avidable at: http://services.parliament.uk/bills/2010-11/localism.html

Unruh, G. C. 2000. Understanding carbon lock-in. *Energy Policy,* 28, 817-830.

USEPA 2010. Green Infrastructure Case Studies: Municipal Policies for Managing Stormwater with Green Infrastructure. United States Environmental Protection Agency. EPA-841-F-10-004. August http://www.epa.gov

Van Den Hurk, B. 2007. New climate change scenarios for the Netherlands. *Water science and technology,* 56, 27-33.

Van Grol R., Walker W.E., Rahman A., de Jong G. 2006. Using a metamodel to analyze sustainable transport policies for europe: The SUMMA project's Fast Simple Model. In: EURO 2006, 21st European Conference on Operational Research, Iceland, 2-5 July 2006.

Van Koningsveld, M. 2004. Plan study Safety for the Province of South Holland. The weak links: the top of North Holland, the Hondsbossche and Pettemer coastal defence. Planstudie Veiligheid voor de Provincie NoordHolland. De zwakke schakels: de kop van NoordHolland, de Hondsbossche en Pettemer zeewering. (In Dutch). Rapport in opdracht van de Provincie NoordHolland. WL rapport Z3725.50.

Van Herk, S., Zevenbergen, C., Ashley, R. & Rijke, J. 2011. Learning and Action Alliances for the integration of flood risk management into urban planning: a new framework from empirical evidence from The Netherlands. *Environmental Science & Policy.*

Veerbeek, W., Ashley, R., Zevenbergen, C., Gersonius, B. & Rijke, J. 2010. Building Adaptive Capacity For Flood Proofing In Urban Areas Through Synergistic Interventions. First International Conference on Sustainable Urbanization (ICSU 2010). Hong Kong, China, 15-17 December.

Ven, F. van de, Van Nieuwkerk, E., Stone, K., Zevenbergen, C., Veerbeek, W., Rijke J. & Van Herk S. 2011. Building the Netherlands climate proof. Urban areas, 1201082-000-VEB-0003, Delft/Utrecht: Deltares/UNESCO-IHE.

VenW 2009. National Water Plan. Available at: http://english.verkeerenwaterstaat.nl/english/topics/water/water_and_the_future/national_water_plan/

VenW 2010. Water Act. Available at: http://www.rijksoverheid.nl/documenten-en-publicaties/rapporten/2010/02/01/water-act.html

Voß, J. P., Newig, J., Kastens, B., Monstadt, J. & Nölting, B. 2007. Steering for Sustainable Development: a typology of problems and strategies with respect to ambivalence, uncertainty and distributed power. *Journal of Environmental Policy & Planning,* 9, 193-212.

Vreugdenhil, B.J., Noortwijk, J.M. van. & De Gooier C. 2000. HYDRA-K: functional design. HYDRA-K: functioneel ontwerp. (In Dutch).

Walker, B. 2010. Preface: a social-ecological perspective of resilience. *Editor and Author/s. Resilience: Interdisciplinary Perspectives on Science and Humanitarianism,* Volume 1.

Walker, B., Carpenter, S., Anderies, J., Abel, N., Cumming, G. S., Janssen, M., Lebel, L., Norberg, J., Peterson, G. D. & Pritchard, R. 2002. Resilience Management in Social-ecological Systems: a Working Hypothesis for a Participatory Approach. *Conservation Ecology,* 6, 14.

Walker, B., Holling, C. S., Carpenter, S. R. & Kinzig, A. 2004. Resilience, Adaptability and Transformability in Social--ecological Systems. *Ecology and Society,* 9, 5.

Walker, B. H., Anderies, J. M., Kinzig, A. P. & Ryan, P. 2006. Exploring resilience in social-ecological systems through comparative studies and theory development: introduction to the special issue. *Ecology and Society,* 11, 12.

Walker, W. E., Haasnoot, M. & Kwakkel, J. H. In prep. Combining adaptation pathways and adaptive policymaking to create adaptive policies for the Rhine Delta.

Walker, W. E., Rahman, S. A. & Cave, J. 2001. Adaptive policies, policy analysis, and policy-making. *European Journal of Operational Research,* 128, 282-289.

Wang, T. 2005. *Real Options" in" Projects and Systems Design-Identification of Options and Solution for Path Dependency.* Phd thesis, Massachusetts Institute of Technology, Cambridge, MA.

Wang, T. & De Neufville, R. 2004. Building Real Options into Physical Systems with Stochastic Mixed-Integer Programming.

Wang, T. & De Neufville, R. 2005. Real options "in" projects. *9th Real Options Annual International Conference, Paris, FR.*

Wiener, N. 1923. Differential space. *Journal of Mathematical Physics,* 2, 131-174.

Willems, P., Arnbjerg-Nielsen, K., Olsson, J. & Nguyen, V. 2012. Climate change impact assessment on urban rainfall extremes and urban drainage: Methods and shortcomings. *Atmospheric research,* 103, 106-118.

WMO. 2009. Guide to Hydrological Practices. Volume II: Management of Water Resources and Application of Hydrological Practices. WMO No. 168. 6th ed. World Meteorological Organisation, Geneva, Switzerland. ISBN 978-92-63-10168-6.

Woodward, M., Gouldby, B., Kapelan, Z., Khu, S. T. & Townend, I. 2011. Real Options in flood risk management decision making. *Journal of Flood Risk Management.*

WWAP. 2012. The United Nations World Water Development Report 4: Managing Water under Uncertainty and Risk. Paris, UNESCO.

Zevenbergen, C., 2007. Adapting to Change: Towards Flood Resilient Cities, UNESCO-IHE, Delft. ISBN: 978-90-73445-18-5.

Zevenbergen, C., Cashman, A., Evelpidou, A., Pasche, E., Garvin, S. & Ashley, R. 2011. *Urban flood management,* CRC Press.

Zevenbergen, C., Veerbeek, W., Gersonius, B. & Van Herk, S. 2008. Challenges in urban flood management: travelling across spatial and temporal scales. *Journal of Flood Risk Management,* 1, 81-88.

Zevenbergen, C., Van Herk, S., Rijke, J., Kabat, P., Bloemen, P., Ashley, R., Speers, A., Gersonius, B. & Veerbeek, W. 2012. Taming global flood disasters. Lessons learned from Dutch experience. Submitted to Natural Hazards.

Zhao, T., Sundararajan, S. & Tseng, C. 2004. Highway development decision-making under uncertainty: A real options approach. *Journal of infrastructure systems,* 10, 23.

Zhao, T. & Tseng, C. 2003. Valuing flexibility in infrastructure expansion. *Journal of infrastructure systems,* 9, 89.

# Appendix A: Technical characteristics

**Table A.1.** Optimal configurations under different approaches to climate adaptation

| Component | Variable | Unit | Existing design | Static robust strategy | Dynamic adaptive strategy | | |
|---|---|---|---|---|---|---|---|
| | | | | | A1 | A2 | A3 |
| J108_000.1 | Diameter | (m) | 0.229 | 0.229 | 0.229 | 0.229 | 0.229 |
| J108_010.1 | Diameter | (m) | 0.305 | 0.305 | 0.305 | 0.305 | 0.305 |
| J10_000.1 | Diameter | (m) | 0.225 | 0.45 | 0.45 | 0.45 | 0.45 |
| J11_000.1 | Diameter | (m) | 0.225 | 0.225 | 0.225 | 0.225 | 0.225 |
| J11_010.1 | Diameter | (m) | 0.225 | 0.225 | 0.225 | 0.225 | 0.225 |
| J12_000.1 | Diameter | (m) | 0.15 | 0.15 | 0.15 | 0.15 | 0.15 |
| J12_010.1 | Diameter | (m) | 0.225 | 0.225 | 0.225 | 0.225 | 0.225 |
| J12_020.1 | Diameter | (m) | 0.225 | 0.225 | 0.225 | 0.225 | 0.225 |
| J13_000.1 | Diameter | (m) | 0.225 | 0.45 | 1.2 | 1.2 | 1.2 |
| J14_000.1 | Diameter | (m) | 0.15 | 0.45 | 0.225 | 0.225 | 0.225 |
| J14_010.1 | Diameter | (m) | 0.15 | 0.45 | 0.45 | 0.45 | 0.45 |
| J14_020.1 | Diameter | (m) | 0.15 | 0.45 | 0.45 | 0.45 | 0.45 |
| J14_030.1 | Diameter | (m) | 0.225 | 0.45 | 0.45 | 0.45 | 0.45 |
| J14_040.1 | Diameter | (m) | 0.225 | 0.3 | 0.3 | 0.3 | 0.3 |
| J16_000.1 | Diameter | (m) | 0.225 | 0.45 | 0.45 | 0.45 | 0.45 |
| J16_030.1 | Diameter | (m) | 0.3 | 0.3 | 0.3 | 0.3 | 0.3 |
| J16_040.1 | Diameter | (m) | 0.225 | 0.225 | 0.225 | 0.225 | 0.225 |
| J16_050.1 | Diameter | (m) | 0.225 | 0.225 | 0.225 | 0.225 | 0.225 |
| J16_070.1 | Diameter | (m) | 0.3 | 0.3 | 0.3 | 0.3 | 0.3 |
| J16_080.1 | Diameter | (m) | 0.375 | 0.375 | 0.375 | 0.375 | 0.375 |
| J16_081.1 | Diameter | (m) | 0.45 | 0.45 | 0.45 | 0.45 | 0.45 |
| J16_090.1 | Diameter | (m) | 0.375 | 0.375 | 0.375 | 0.375 | 0.375 |
| J16_100.1 | Diameter | (m) | 0.3 | 0.3 | 0.45 | 0.45 | 0.45 |
| J16_110.1 | Diameter | (m) | 0.6 | 0.6 | 0.6 | 0.6 | 0.6 |
| J16_120.1 | Diameter | (m) | 0.45 | 0.45 | 0.45 | 0.45 | 0.45 |
| J16_130.1 | Diameter | (m) | 0.3 | 0.45 | 0.6 | 0.6 | 0.6 |
| J16_140.1 | Diameter | (m) | 0.375 | 0.375 | 0.375 | 0.375 | 0.375 |
| J16_150.1 | Diameter | (m) | 0.45 | 0.45 | 0.45 | 0.45 | 0.45 |
| J1_000.1 | Diameter | (m) | 0.225 | 1.2 | 0.225 | 0.225 | 0.225 |
| J1_010.1 | Diameter | (m) | 0.225 | 0.225 | 0.45 | 0.45 | 0.45 |
| J1_020.1 | Diameter | (m) | 0.225 | 0.45 | 0.45 | 0.45 | 0.45 |
| J1_030.1 | Diameter | (m) | 0.225 | 0.225 | 0.225 | 0.225 | 0.225 |
| J1_040.1 | Diameter | (m) | 0.225 | 0.225 | 0.225 | 0.225 | 0.225 |
| J1_045.1 | Diameter | (m) | 0.3 | 0.3 | 0.3 | 0.3 | 0.3 |
| J1_050.1 | Diameter | (m) | 0.3 | 0.45 | 0.45 | 0.45 | 0.45 |
| J1_060.1 | Diameter | (m) | 0.45 | 0.45 | 0.45 | 0.45 | 0.45 |
| J1_063.1 | Diameter | (m) | 0.45 | 0.45 | 0.45 | 0.45 | 0.45 |
| J1_067.1 | Diameter | (m) | 0.45 | 0.45 | 0.45 | 0.45 | 0.45 |
| J1_068.1 | Diameter | (m) | 0.389 | 0.389 | 0.389 | 0.389 | 0.389 |
| J1_069.1 | Diameter | (m) | 0.45 | 0.45 | 0.45 | 0.45 | 0.45 |
| J1_080.1 | Diameter | (m) | 1.27 | 1.27 | 1.27 | 1.27 | 1.27 |
| J1_081.1 | Diameter | (m) | 1.27 | 1.27 | 1.27 | 1.27 | 1.27 |
| J1_082.1 | Diameter | (m) | 1.27 | 1.27 | 1.27 | 1.27 | 1.27 |

| | | | | | | | |
|---|---|---|---|---|---|---|---|
| J1_090.1 | Diameter | (m) | 0.45 | 0.45 | 0.6 | 0.6 | 0.6 |
| J1_091.1 | Diameter | (m) | 0.3 | 0.45 | 0.45 | 0.45 | 0.45 |
| J1_100.1 | Diameter | (m) | 0.45 | 0.6 | 0.6 | 0.6 | 0.6 |
| J1_110.1 | Diameter | (m) | 0.45 | 1.2 | 0.9 | 0.9 | 0.9 |
| J1_115.1 | Diameter | (m) | 0.61 | 0.61 | 0.61 | 0.61 | 0.61 |
| J1_130.1 | Diameter | (m) | 0.61 | 0.61 | 0.9 | 0.9 | 0.9 |
| J1_140.1 | Diameter | (m) | 0.61 | 0.61 | 1.2 | 1.2 | 1.2 |
| J1_150.1 | Diameter | (m) | 0.61 | 0.61 | 0.61 | 0.61 | 0.61 |
| J1_170.1 | Diameter | (m) | 0.61 | 0.61 | 0.9 | 0.9 | 0.9 |
| J1_180.1 | Diameter | (m) | 0.61 | 0.9 | 0.61 | 0.61 | 0.61 |
| J1_200.1 | Diameter | (m) | 0.61 | 0.9 | 0.9 | 0.9 | 0.9 |
| J1_210.1 | Diameter | (m) | 0.61 | 1.2 | 0.9 | 0.9 | 0.9 |
| J20_000.1 | Diameter | (m) | 0.15 | 0.225 | 0.225 | 0.225 | 0.225 |
| J21_000.1 | Diameter | (m) | 0.15 | 0.15 | 0.15 | 0.15 | 0.15 |
| J21_010.1 | Diameter | (m) | 0.15 | 0.45 | 0.225 | 0.225 | 0.225 |
| J21_020.1 | Diameter | (m) | 0.15 | 0.9 | 0.225 | 0.225 | 0.225 |
| J21_030.1 | Diameter | (m) | 0.225 | 0.225 | 0.45 | 0.45 | 0.45 |
| J22_000.1 | Diameter | (m) | 0.15 | 0.15 | 0.9 | 0.9 | 0.9 |
| J24_000.1 | Diameter | (m) | 0.15 | 0.15 | 0.225 | 0.225 | 0.225 |
| J26_000.1 | Diameter | (m) | 0.3 | 0.3 | 0.3 | 0.3 | 0.3 |
| J27_020.1 | Diameter | (m) | 0.225 | 0.45 | 0.3 | 0.3 | 0.3 |
| J28_000.1 | Diameter | (m) | 0.225 | 0.225 | 0.225 | 0.225 | 0.225 |
| J29_000.1 | Diameter | (m) | 0.225 | 0.45 | 0.45 | 0.45 | 0.45 |
| J29_010.1 | Diameter | (m) | 0.225 | 0.45 | 0.45 | 0.45 | 0.45 |
| J29_020.1 | Diameter | (m) | 0.225 | 0.45 | 0.45 | 0.45 | 0.45 |
| J29_030.1 | Diameter | (m) | 0.225 | 0.6 | 0.45 | 0.45 | 0.45 |
| J2_000.1 | Diameter | (m) | 0.225 | 0.45 | 0.45 | 0.45 | 0.45 |
| J30_000.1 | Diameter | (m) | 0.225 | 0.45 | 0.45 | 0.45 | 0.45 |
| J3_000.1 | Diameter | (m) | 0.225 | 0.45 | 0.45 | 0.45 | 0.45 |
| J5_000.1 | Diameter | (m) | 0.225 | 1.2 | 0.45 | 0.45 | 0.45 |
| J6_000.1 | Diameter | (m) | 0.225 | 0.225 | 0.225 | 0.225 | 0.225 |
| J7_000.1 | Diameter | (m) | 0.225 | 0.45 | 0.45 | 0.45 | 0.45 |
| J8_000.1 | Diameter | (m) | 0.15 | 0.225 | 0.225 | 0.225 | 0.225 |
| J8_010.1 | Diameter | (m) | 0.375 | 0.375 | 0.375 | 0.375 | 0.375 |
| J8_020.1 | Diameter | (m) | 0.375 | 0.9 | 0.9 | 0.9 | 0.9 |
| J8_050.1 | Diameter | (m) | 0.5 | 0.5 | 0.5 | 0.5 | 0.5 |
| J8_060.1 | Diameter | (m) | 0.375 | 0.375 | 0.375 | 0.375 | 0.375 |
| J8_070.1 | Diameter | (m) | 0.375 | 0.375 | 0.375 | 0.375 | 0.375 |
| Virtual link | Diameter | (m) | 0.61 | 0.61 | 0.61 | 0.61 | 0.61 |
| J16_060.1 | Diameter | (m) | 0.3 | 0.3 | 0.3 | 0.3 | 0.3 |
| J1_070.1 | Diameter | (m) | 0.9 | 0.9 | 0.9 | 0.9 | 0.9 |
| J1 | Diameter | (m) | 0.45 | 0.45 | 0.45 | 0.45 | 0.45 |
| S1_080 | Area | (m$^2$) | 0 | 250 | 250 | 250 | 250 |
| S1_060 | Area | (m$^2$) | 0 | 250 | 250 | 250 | 250 |
| S1_115 | Area | (m$^2$) | 0 | 1000 | 500 | 1000 | 1000 |
| S6_000 | Area | (m$^2$) | 0 | 500 | 500 | 1000 | 2000 |
| S108_000 | Area | (m$^2$) | 0 | 1500 | 0 | 500 | 2000 |

| | | | | | | | |
|---|---|---|---|---|---|---|---|
| S16_040 | Area | (m²) | 0 | 500 | 500 | 1000 | 2000 |
| S16_080 | Area | (m²) | 0 | 0 | 0 | 0 | 750 |
| S8_070 | Area | (m²) | 0 | 1750 | 0 | 500 | 2000 |
| S1_030 | Area | (m²) | 0 | 1000 | 500 | 1000 | 1000 |
| C1001 | % Roofs | (%) | 0 | 0 | 0 | 0 | 0 |
| C1002 | % Roofs | (%) | 0 | 45 | 0 | 45 | 60 |
| C1003 | % Roofs | (%) | 0 | 0 | 0 | 0 | 0 |
| C1004 | % Roofs | (%) | 0 | 0 | 0 | 0 | 0 |
| C1005 | % Roofs | (%) | 0 | 45 | 0 | 15 | 60 |
| C1006 | % Roofs | (%) | 0 | 45 | 0 | 15 | 60 |
| C1007 | % Roofs | (%) | 0 | 0 | 0 | 15 | 60 |
| C1008 | % Roofs | (%) | 0 | 0 | 0 | 15 | 60 |
| C1009 | % Roofs | (%) | 0 | 0 | 0 | 0 | 0 |
| C1010 | % Roofs | (%) | 0 | 0 | 0 | 0 | 0 |
| C1011 | % Roofs | (%) | 0 | 30 | 0 | 0 | 30 |
| C1012 | % Roofs | (%) | 0 | 0 | 0 | 0 | 0 |
| C1013 | % Roofs | (%) | 0 | 0 | 0 | 0 | 0 |
| C1014 | % Roofs | (%) | 0 | 0 | 0 | 0 | 0 |
| C1015 | % Roofs | (%) | 0 | 0 | 0 | 30 | 30 |
| | Cost | (M£) | | 2.13 | 1.66 | 1.94 | 2.50 |

# Acknowledgements

My sincere gratitude goes to my two supervisors, Prof. Chris Zevenbergen and Prof. Richard Ashley, for their scientific guidance, support and motivation throughout the 4-year period of research. Chris, your views on resilience and system dynamics have been a key source of inspiration for this research. Richard, thank you for sharing your broad expertise on water and flood risk management with me, and for providing critical feedback on this thesis and the related papers. I would also like to thank my daily supervisor, Dr. Assela Pathirana, for his guidance on flood risk modelling and computer programming. Your enthusiasm for scientific research and patience with researchers have been very inspirational. I thank Prof. Kala Vairavamoorthy for encouraging me to pursue a PhD at UNESCO-IHE.

This research has been carried out as a part of the EU Interreg IVB project MARE, which has developed and demonstrated best practices for "Managing Adaptive REsponses to changing flood risk in the North Sea Region". I would like to thank the EU for funding this project. Furthermore, I am thankful for the funding of Dura Vermeer Group, the Living with Water programme, and the Delta Programme.

This research would not have been possible without the support and cooperation of the participants from the MARE LAA Dordrecht. Special thanks are due to Ellen Kelder (City of Dordrecht) for championing the LAA and shaping its innovation. I want to acknowledge the help of Jannekee van Herreveld (WSHD), Jacob Luijendijk and Arjan van der Lee (Tauw) in applying the integrated urban drainage model of Dordrecht. I am also grateful to Prof. Simon Tait and Richard Newman (PWG) for providing the urban drainage model of West Garforth. I owe much to the co-authors of the papers included in this thesis: Ad Jeuken, Teun van Morselt, Fauzy Nasruddin, and Leo van Nieuwenhuijzen.

The quality of this research has greatly benefited from the discussions with many fellow researchers (e.g., Cost C22), the peer-review comments on the papers, and the suggestions for improvement of the draft thesis by the members of the doctoral committee.

Many thanks go my colleagues at UNESCO-IHE for making this PhD research a more enjoyable experience, in particular to my direct colleagues at the former UWS Department and the WSE Department. Of course, this PhD experience would not been the same without the stimulating company and friendship of my fellow PhD candidates and/or colleagues at FRG: Sebastiaan van Herk, William

Veerbeek, Jeroen Rijke, Kim Anema, Taneha Bacchin, Koen Olthuis, Ellen Brandenburg, Natasa Manojlovic and Edwin van Son.

Last but not least, I want to thank to my family and friends for providing welcome distraction from my PhD research. Menzo and Evelien, thank you for believing in me, and for your love and support. Anja, I am fortunate to have such a nice and caring sister. And Victoria, thank you for always being there, to support me in difficult times and to share happiness in good times. You are the best part of my life.

# Curriculum Vitae

Dec. 18, 1980:     Born in Alkmaar as BERRY GERSONIUS.

## Education

1999 – 2002:     BSc in Civil Engineering, Delft University of Technology, Delft, The Netherlands.

2002 – 2005:     MSc in Civil Engineering, Delft University of Technology, Delft, The Netherlands. Specialisation in Water Infrastructure Planning.

## Employment record

2006 – 2010:     Lecturer Sustainable Urban Infrastructure Systems, Urban Water and Sanitation Department, UNESCO-IHE, Delft, The Netherlands.

2010 – present:     Lecturer Urban Flood Resilience, Water Science and Engineering Department, UNESCO-IHE, Delft, The Netherlands.

## Experience record

2006 – 2008:     Member LmW Project: Urban Flood Management Dordrecht.

2007 – 2008:     Member CcS Project: COM23 Water Robust Building.

2007 – 2009:     Member UNESCO-IHP VI Project: Integrated Urban Water System Interactions.

2007 – 2010:     Member EU COST Action C22: Urban Flood Magement.

2008:     Member Advisory Committee: Procedures en wet- en regelgeving voor klimaatbestendig bouwen.

2009 – present:     Member EU Interreg IVB Project: MARE.

2010:     Member MTEC Course: Water Framework Directive and Flood Directive.

2010:     Member LmW Project: Wielwijk Klimaatbestendig.

2010:     Member LmW Project: Delft Spetterstad.

2010:     Member Project: Reële Opties in Kustbeheer

2010 – present:     Member EU FP7 Project: FloodProbe.

2010 – present:     Member Advisory Committee RAAK project: Anticiperen op extreme neerslag in de stad.

2011:     Member FWR Project: ROCK Surface Water Management and Urban Green Infrastructure.

2011:     Member Advanced Course / Workshop: Adaptive Waterscapes - Designing Water Resilient Urban Environments.

| 2011 – present: | Member CcS Project: HSRR 3.1 Adaptieve strategieën in het buitendijks gebied van Rotterdam. |
|---|---|
| 2011 – present: | Member Project: PCT-based flood-proofing of road infrastructure. |
| 2011 – present: | Member DP RD Working Group: Anders Omgaan Met Water. |
| 2011 – present: | Member Advisory Committee STOWA Project: Afwegingskader Meerlaagsveiligheid. |
| 2011 – present: | Member Think Tank: Delft Urban Water. |
| 2012 – present: | Member Knowledge Network: Adaptief Delta Management. |
| 2012 – present: | Member WOTRO Integrated Project: Dynamic Deltas. |
| 2012 – present: | Member CRC for Water Sensitive Cities Project: B5 Socio-Technical Flood Resilience. |

# Publications

## Peer-reviewed journals

Ashley, R., Blanskby, J., Newman, R., Gersonius, B., Poole, A., Lindley, G., Smith, S., Ogden, S. & Nowell, R. 2012. Learning and Action Alliances to build capacity for flood resilience. *Journal of Flood Risk Management,* 5, 14-22.

Gersonius, B., Ashley, R., Jeuken, A., Pathirana, A. & Zevenbergen, C. 2012. Accounting for uncertainty and flexibility in flood risk management: comparing Real In Options and Adaptation Tipping Points. *Journal of Flood Risk Management,* under review.

Gersonius, B., Ashley, R., Pathirana, A. & Zevenbergen, C. 2012. Adaptation of flood risk infrastructure for climate resilience. *Proceedings of the ICE - Civil Engineering,* accepted.

Gersonius, B., Ashley, R., Pathirana, A. & Zevenbergen, C. 2012. Climate change uncertainty: building flexibility into water and flood risk infrastructure. *Climatic Change,* accepted.

Gersonius, B., Ashley, R. & Zevenbergen, C. 2012. The identity approach for assessing socio-technical resilience to climate change: example of flood risk man-agement for the Island of Dordrecht. *Natural Hazards and Earth System Sciences,* under review.

Gersonius, B., Nasruddin, F., Ashley, R., Jeuken, A., Pathirana, A. & Zevenbergen, C. 2012. Developing the evidence base for mainstreaming adaptation of stormwater systems to climate change. *Water Research,* accepted.

Pathirana, A., Gersonius, B. & Vairavamoorthy, K. 2012. Responding Responsibly to Global Change in Water Infrastructure Planning. *Urban Water,* submitted.

Pathirana, A., Gersonius, B. & Radhakrishnan, M. 2012. Web 2.0 collaboration tools to support student research in hydrology–an opinion. *Hydrology and Earth System Sciences Discussions,* 9, 2541-2567.

Delelegn, S., Pathirana, A., Gersonius, B., Adeogun, A. & Vairavamoorthy, K. 2011. Multi-objective optimisation of cost-benefit of urban flood management using a 1 D 2 D coupled model. *Water science and technology,* 63, 1054.

Gersonius, B., Morselt, T., Van Nieuwenhuijzen, L., Ashley, R. & Zevenbergen, C. 2011. How the Failure to Account for Flexibility in the Economic Analysis of Flood Risk and Coastal Management Strategies Can Result in Maladaptive Decisions. *Journal of Waterway, Port, Coastal, and Ocean Engineering,* 1, 96.

Ven, F., Gersonius, B., Graaf, R., Luijendijk, E. & Zevenbergen, C. 2011. Creating water robust urban environments in the. *Journal of Flood Risk Management,* 4, 273-280.

Gersonius, B., Ashley, R., Pathirana, A. & Zevenbergen, C. 2010. Managing the flooding system's resiliency to climate change. *Proceedings of the Institution of Civil Engineers - Engineering Sustainability,* 163, 15-22.

Maharjan, M., Pathirana, A., Gersonius, B. & Vairavamoorthy, K. 2009. Staged cost optimization of urban storm drainage systems based on hydraulic performance in a changing environment. *Hydrology and Earth System Sciences,* 13, 481-489.

Jumadar, A., Pathirana, A., Gersonius, B. & Zevenbergen, C. 2008. Incorporating infiltration modelling in urban flood management. *Hydrology and Earth System Sciences Discussions,* 5.

Pathirana, A., Tsegaye, S., Gersonius, B. & Vairavamoorthy, K. 2008. A simple 2-D inundation model for incorporating flood damage in urban drainage planning. *Hydrology and Earth System Sciences Discussions,* 5, 3061-3097.

Zevenbergen, C., Veerbeek, W., Gersonius, B. & Van Herk, S. 2008. Challenges in urban flood management: travelling across spatial and temporal scales. *Journal of Flood Risk Management,* 1, 81-88.

## Authorship

Ashley R., Nowell R., Gersonius B. & Walker L. 2011. *Surface Water Management and Urban Green Infrastructure – A review of potential benefits and UK and international practices.* FR/R0014. Foundation for Water Research, Allen House, The Listons, Liston Road, Marlow, Bucks. SL7 1FD UK.

Ven, F. van de, Luyendijk, E., Gunst M. de, Tromp, E., Schilt, M., Krol, L., Gersonius, B., Vlaming, C., Valkenburg, L. & Peeters, R. 2009. *Waterrobuust bouwen. De kracht van kwetsbaarheid in een duurzaam ontwerp.* KvR report 016/2010. Beter Bouw- en Woonrijp Maken/SBR, Rotterdam. ISBN 978-90-8815-017-3.

## Book contributions

Gersonius, B. 2010. *C. Zevenbergen R. Ashley, S. Garvin, E. Pasche, N. Evelpidou & A. Cashman, Urban Flood Management.*

Gersonius, B., Zevenbergen, C. & Herk, S. 2007. Managing flood risk in the urban environment: linking spatial planning, risk assessment, communication and policy. *C. Pahl Wostl, P. Kabat, & J. Möltgen, Adaptive and Integrated Water Management: coping with complexity and uncertainty,* 263-275.

Zevenbergen, C. & Gersonius, B. 2007. Challenges in Urban Flood Management. *R. Ashley, S. Garvin, E. Pasche, A. Vassilopoulos, & C. Zevenbergen, Advances in Urban Flood Management,* 1-12.

Zevenbergen, C., Gersonius, B., Puyan, N. & Van Herk, S. 2007. Economic feasibility study of flood proofing domestic dwellings. *R. Ashley, S. Garvin, E. Pasche, A. Vassilopoulos, & C. Zevenbergen, Advances in Urban Flood Management*, 299-318.

## Conference proceedings

Gersonius, B., Ashley, R., Blanksby, J. Walker, L., Richter, S., Zeller, S. & Stone, K. 2012. Water Sensitive Urban Design as an essential component of adapting water systems to cope with flood risks. *WSUD2012*.

Walker, L., Ashley, R., Nowell, R., Gersonius, B. & Evans. T. 2012. Surface water management and urban green infrastructure in the UK: A review of benefits and challenges. *WSUD 2012*.

Gersonius, B., Ashley, R., Jeuken, A., Pathirana, A. & Zevenbergen, C. 2011. Accounting for climate change in urban drainage and flooding: contrasting alternative approaches to devising adaptive strategies. *12ICUD*.

Gersonius, B., Veerbeek, W., Subhan, A., Stone, K. & Zevenbergen, C. 2011. Toward a More Flood Resilient Urban Environment: The Dutch Multi-level Safety Approach to Flood Risk Management. *Resilient Cities*, 273-282.

Ashley, R., Faram, M., Chatfield, P., Gersonius, B. & Andoh, R. 2010. Appropriate Drainage Systems for a Changing Climate in the Water Sensitive City. *Low Impact Development 2010: Redefining Water in the City*, 864-877

Gersonius, B., Ashley, R., Jeuken, A., Nasruddin, F., Pathirana, A. & Zevenbergen, C. 2010. A resilience perspective to water risk management: case-study application of the adaptation tipping point method. *EGU General Assembly 2010*, 1627.

Gersonius, B., Ashley, R., Pathirana, A. & Zevenbergen, C. 2009. A conceptual basis for the application of resilience with respect to flooding systems: an overview. *Road Map Towards a Flood Resilient Urban Environment*.

Gersonius, B. 2008. Can resilience support integrated approaches to urban drainage management? *11ICUD*.

Gersonius, B., Tucci, C.E.M., & Zevenbergen, C. 2008. Critical factors for integrated drainage management plans: experiences of Porto Alegre, Brazil. *11ICUD*.

Gersonius, B., Zevenbergen, C., Puyan, N. & Billah, M. 2008. Efficiency of private flood proofing of new buildings–Adapted redevelopment of a floodplain in the Netherlands. *WIT Transactions on Ecology and the Environment*, 118, 247-259.

## National publications

Gersonius, B., Zevenbergen, C., Jonkman, B., Kanning, W. & Ter Horst, W. 2012. Deelprogramma Veiligheid. *H2O*.

Zevenbergen, C., Gersonius, B., Penning, E. & Bakker, E.F. 2004. Duurzaam ruimtegebruik: Drijvende Kassen. *Civiele Techniek*.

T - #0097 - 071024 - C42 - 244/170/10 - PB - 9780415624855 - Gloss Lamination